T0325849

Modelling Freight Transport

Modelling Freight Transport

Edited by

Lóránt Tavasszy
Delft University of Technology
The Netherlands

Gerard de Jong
Institute for Transport Studies
University of Leeds, UK

AMSTERDAM • BOSTON • HEIDELBERG • LONDON • NEW YORK • OXFORD
PARIS • SAN DIEGO • SAN FRANCISCO • SINGAPORE • SYDNEY • TOKYO

Elsevier
32 Jamestown Road, London NW1 7BY
225 Wyman Street, Waltham, MA 02451, USA

First Edition 2014

British Library Cataloguing-in-Publication Data
A catalogue record for this book is available from the British Library

Library of Congress Cataloging-in-Publication Data
A catalog record for this book is available from the Library of Congress

ISBN: 978-0-12-410400-6

For information on all Elsevier publications
visit our website at store.elsevier.com

Working together
to grow libraries in
developing countries

www.elsevier.com • www.bookaid.org

Contents

List of Contributors   xi

1 Introduction   1
Lóránt Tavasszy and Gerard de Jong

  1.1 Background and Objectives   1
  1.2 Conceptual Framework for Freight Decisions   2
      1.2.1 Goods: Production, Consumption and Trade   3
      1.2.2 Inventory Networks   4
      1.2.3 Transport Organisation   4
  1.3 Freight Models – Theoretical Perspective of the Book   6
  1.4 Freight Models – Practical Perspectives Addressed   9
  1.5 Organisation of This Volume   10
  References   11

2 Modelling Inter-Regional Freight Demand with Input–Output, Gravity and SCGE Methodologies   13
Olga Ivanova

  2.1 Introduction   13
  2.2 State-of-the-Art in of Inter-Regional Freight
      Demand Modelling   16
  2.3 Forecasting Inter-Regional Trade Using Gravity   19
      2.3.1 The History of Gravity Model   19
      2.3.2 Theoretical Foundations of Gravity Equation   20
  2.4 Multi-Regional I/O Framework   21
      2.4.1 Multi-Regional IO Models with Fixed
              Trade Coefficients   21
      2.4.2 Introduction of Variable Trade Coefficients
              Using Random Utility Theory   22
      2.4.3 Overview of Empirical Inter-Regional IO Models 23
  2.5 Spatial General Equilibrium Models and NEG Effects   25
      2.5.1 Overview of the Methodology   25

2.5.2 Calibration of SCGE Models   26
2.5.3 Simple Mathematical Formulation   27
2.5.4 Including the Spatial Dimension   29
2.5.5 Capital and Labour Mobility   30
2.5.6 Agglomeration/Dispersion Forces   30
2.5.7 Review of the Empirical SCGE Models   31
2.5.8 The RAEM Family of SCGE Models   35
2.6 Conclusions and Ideas for Further Research   37
References   39
Suggestions for Further Reading   40

3 **Freight Generation and Freight Trip Generation Models**   **43**
José Holguín-Veras, Miguel Jaller, Ivan Sánchez-Díaz,
Shama Campbell and Catherine T. Lawson

3.1 Introduction   43
3.2 Literature Review   44
    3.2.1 FTG Models   44
    3.2.2 FG Models   47
3.3 Logistical Interpretation of FG/FTG   47
3.4 Factors to Take into Account when Estimating or Applying
     FG and FTG   50
    3.4.1 Classification Systems   51
    3.4.2 Level of Aggregation   52
    3.4.3 Aggregation Procedures   52
3.5 Case Study: FTG in the New York City Metropolitan Area 53
    3.5.1 Transferability of FG and FTG Models   55
    3.5.2 Spatial Effects on Freight Trips Attraction   57
3.6 Conclusion   57
References   60

4 **Distribution Structures**   **65**
Hanno Friedrich, Lóránt Tavasszy and Igor Davydenko

4.1 Introduction   65
4.2 The Micro Level   66
    4.2.1 Drivers and Their Developments   68
    4.2.2 Micro-Level Normative Models   70
    4.2.3 Applicability for Descriptive Purposes   76

4.3 From Micro to Macro Level   77
    4.3.1 Challenges in Aggregate Models   77
    4.3.2 Aggregate Modelling of Inventory Structures   79
4.4 Conclusion   85
References   86

5 Inventory Theory and Freight Transport Modelling   **89**
François Combes

5.1 Inventory Theory   89
    5.1.1 The Economic Order Quantity Model   90
    5.1.2 Extensions of the EOQ Model   92
    5.1.3 Models of Optimal Shipment Size with
          Uncertain Demand   95
5.2 Microeconomics of Logistics and Freight Transport   97
    5.2.1 The TLC Function   97
    5.2.2 The TLC in the EOQ Model: A Simple Freight
          Mode Choice Model with Logistics   98
    5.2.3 The TLC in the Context of a Dynamic Model:
          A Partial Theory of the Value of Time and
          Value of Reliability in Freight Transport   100
    5.2.4 The Inventory Theoretic Model of Freight
          Transport Demand of W. J. Baumol and H. D. Vinod   101
    5.2.5 The Structure of Freight Transport Costs   102
5.3 Databases   103
    5.3.1 Commodity Flow Surveys   103
    5.3.2 The ECHO Shipper Survey   104
    5.3.3 Inventory Theory and the Need for Adequate Data   104
5.4 The Econometrics of Freight Mode Choice and
    Shipment Size   105
    5.4.1 Exploratory Analyses   105
    5.4.2 Structural Analyses   106
5.5 Perspectives for Simulation   111
References   112

6 Mode Choice Models   **117**
Gerard de Jong

6.1 Introduction   117

6.1.1 Mode Choice at Different Spatial Levels   **117**
6.1.2 Relevance of Modal Split   **118**
6.1.3 Dependent and Independent Variables   **118**
6.1.4 Disaggregate and Aggregate Mode Choice Models   **119**
6.1.5 Closely Related Choices   **119**
6.2 The Disaggregate Mode Choice Theory   **121**
6.2.1 Cost Functions and Utility Functions   **121**
6.2.2 Different Distributional Assumptions
      Lead to Different Discrete Choice Models   **124**
6.2.3 Non-RUM Models   **129**
6.2.4 Interaction Between Agents in Freight Transport   **130**
6.3 Practical Examples   **132**
6.3.1 Aggregate Mode Choice Models   **132**
6.3.2 Disaggregate Mode Choice Models   **134**
6.3.3 Joint Models for Mode Choice and Related Choices   **135**
References   **137**

7 Vehicle-Trip Estimation Models   **143**
José Holguín-Veras, Carlos González-Calderón,
Iván Sánchez-Díaz, Miguel Jaller and Shama Campbell

7.1 Introduction   **143**
7.2 Estimation of Loaded Trips   **146**
7.3 Estimation of Empty Trips   **149**
7.3.1 Commodity-Based Empty Trip Models   **150**
7.3.2 Generalised NVE Models   **151**
7.3.3 Tour Based Empty Trip Models   **152**
7.3.4 Parameter Estimation Procedures   **155**
7.3.5 Empirical Evidence   **158**
7.4 Concluding Remarks   **159**
References   **160**

8 Urban Freight Models   **163**
Antonio Comi, Rick Donnelly and Francesco Russo

8.1 Introduction   **163**
8.2 Push Models of Urban Freight   **166**
8.2.1 Classical Urban Truck Models   **166**
8.2.2 Tour-Based Extensions to Classical Models   **168**

8.2.3 Urban Input−Output Data and Models 169
8.2.4 Firmographic Models and Business Metrics 171
8.2.5 The Supply Chain Context of Urban Freight 172
8.2.6 Simulation Modelling Frameworks 174
8.3 Pull Models for Urban Freight 176
8.3.1 Retailer's Standpoint 177
8.3.2 Final Business Standpoint 181
8.3.3 End-Consumers' Standpoint 182
8.3.4 The Overall Modelling Framework 186
8.4 Emerging Modelling Approaches 187
8.5 Conclusions 190
Acknowledgements 191
References 191

9 **Freight Service Valuation and Elasticities** 201
Gerard de Jong

9.1 Introduction 201
9.2 Freight Service Valuation 202
9.2.1 Use of Models Versus Calculation of Factor Cost 202
9.2.2 Different Data and Discrete Choice Models
for Freight Service Valuation 203
9.2.3 Outcomes for the Value of Transport Time 209
9.2.4 Outcomes for the Value of Transport Time Reliability 213
9.2.5 Other Freight Service Values 216
9.3 Freight Transport Elasticities 216
9.3.1 Derivation of Elasticities from Transport Models 216
9.3.2 Classification of Response Mechanisms 219
9.3.3 Outcome Range for Price (Cost) Elasticities 221
9.3.4 Consistent Cost and Time Elasticities of the
Modal Split 224
References 225

10 **Data Availability and Model Form** 229
Lóránt Tavasszy and Gerard de Jong

10.1 Introduction 229
10.2 Overview of Different Data Sources for Freight
Transport Modelling 229

10.2.1   International Trade Statistics   229
10.2.2   National Accounts   231
10.2.3   Transport Statistics   232
10.2.4   Shipper Surveys   233
10.2.5   Specific Project-Based Interview Data
         (Especially Stated Preference Data)   234
10.2.6   Consignment Bills and RFIDs   234
10.2.7   Traffic Count Data   235
10.2.8   Transport Safety Inspection Data   235
10.2.9   Network Data   235
10.2.10 Cost Functions   235
10.2.11 Terminal Data   236
10.3   Which Data Sources Can Be Used in Which
       Type of Model?   236
10.4   Discussion on Data Availability and Model Form   236
10.5   Dealing with Data Limitations Through Estimation   240
10.6   Concluding Remarks   242
References   243

11 Comprehensive Versus Simplified Models   **245**
Lóránt Tavasszy and Gerard de Jong

11.1   Introduction   245
11.2   High- and Low-Resolution Models   246
11.3   Model Objectives and Policy Questions and Their
       Impact on Model Form   248
11.4   Approaches for Simplification   250
       11.4.1   Simplification by Omission of Sub-Models   251
       11.4.2   Simplification by Integration   253
       11.4.3   Simplification by Reduced Data Need   255
11.5   Concluding Remarks on Comprehensive Versus
       Simplified Models   255
References   256

List of Contributors

Shama Campbell Center for Infrastructure, Transportation, and the Environment, and the VREF's Center of Excellence for Sustainable Urban Freight Systems, Department of Civil and Environmental Engineering, Rensselaer Polytechnic Institute, Troy, NY, USA

François Combes Université Paris-Est, LVMT, UMR T9403 ENPC IFSTTAR UMLV, Marne-La-Vallée, Cedex, France

Antonio Comi Department of Enterprise Engineering, University of Rome "Tor Vergata", Rome, Italy

Carlos González-Calderón Center for Infrastructure, Transportation, and the Environment, and the VREF's Center of Excellence for Sustainable Urban Freight Systems, Department of Civil and Environmental Engineering, Rensselaer Polytechnic Institute, Troy, NY, USA

Igor Davydenko TNO, Delft and Delft University of Technology, The Netherlands

Gerard de Jong Institute for Transport Studies, University of Leeds, UK; Significance BV, The Hague, The Netherlands; and Centre for Transport Studies, VTI/KTH, Stockholm, Sweden

Rick Donnelly Parsons Brinckerhoff, Inc., Albuquerque, NM, USA

Hanno Friedrich Technical University Darmstadt, Germany

José Holguín-Veras Center for Infrastructure, Transportation, and the Environment, and the VREF's Center of Excellence for Sustainable Urban Freight Systems, Department of Civil and Environmental Engineering, Rensselaer Polytechnic Institute, Troy, NY, USA

Olga Ivanova Dutch Organization for Applied Scientific Research (TNO), Strategy and Policy, The Netherlands

Miguel Jaller Center for Infrastructure, Transportation, and the Environment, and the VREF's Center of Excellence for Sustainable Urban Freight Systems, Department of Civil and Environmental Engineering, Rensselaer Polytechnic Institute, Troy, NY, USA

Catherine T. Lawson Department of Geography and Planning University at Albany, Albany, NY, USA

Francesco Russo Facoltà di Ingegneria, Università Mediterranea di Reggio Calabria, Reggio Calabria (RC), Italy

Iván Sánchez-Díaz Center for Infrastructure, Transportation, and the Environment, and the VREF's Center of Excellence for Sustainable Urban Freight Systems, Department of Civil and Environmental Engineering, Rensselaer Polytechnic Institute, Troy, NY, USA

Lóránt Tavasszy TNO, Delft and Delft University of Technology, The Netherlands

1 Introduction

Lóránt Tavasszy[a] and Gerard de Jong[b]

[a]TNO, Delft and Delft University of Technology, The Netherlands
[b]Institute for Transport Studies, University of Leeds, UK; Significance BV,
The Hague, The Netherlands; and Centre for Transport Studies, VTI/KTH,
Stockholm, Sweden

1.1 Background and Objectives

Freight transport is an essential part of our economy as it fulfils a unique service within supply chains, bridging the distances between spatially separated places of supply and demand. As is the case with passenger transport, accessibility of places for freight is vital or the economic development of society. Freight transport flows have been growing continuously in the past, due to an increase in population, falling trade barriers and declining transport costs. In addition, the growth of freight flows is propelled by increasing consumption levels and the customisation of products and services. This growth has been facilitated by major infrastructure extensions including roads, railways, waterways, ports and storage and transhipment activities. In recent decades, however, freight flows have also become an area of concern to public policy in a very different way. This relates to the aim of protecting the environment from the negative side effects (e.g. health-related local emissions, greenhouse gas emissions, traffic accidents) of freight transport growth, as is also the case in passenger transport.

Our final goal with this book is to disseminate better tools to evaluate freight transport policies. We give a concise description of the state of the art of mathematical models of the freight transport system, focussing mainly on areas where it deviates from passenger transport models. Such mathematical models can support freight transport policy design in different ways, including:

- Description of flows in a base year and explanation of the drivers of freight transport,
- Forecasting of flows or exploration of alternative futures,
- Performance assessment of freight systems, e.g. for cost-benefit analysis,
- Design and optimisation of freight systems.

Freight transport models have been around in transport research since the early 1960s, and appeared more or less in parallel with passenger transport models. Despite the fact that the underlying economic and statistical theories are similar and date back much further, the application and development of freight models

Modelling Freight Transport. DOI: http://dx.doi.org/10.1016/B978-0-12-410400-6.00001-X

took place much more slowly. This was partly due to the above-mentioned late appearance of freight transport as a major area of public policy. More importantly, perhaps, it was the lack of data, or appropriate behavioural or applied economic theory that distinguished the drivers of freight transport from those of passenger transport. In the 1970s, freight was largely treated by the research community in a simplistic way, as a separate class of passengers, leaning on the same theoretical underpinning and the same applied models. At present, in the community of practitioners, it largely still is. In the meantime, however, new freight-related disciplines have emerged, like logistics and supply chain management, aiming at improving firm logistics, according to principles of optimisation of service levels and minimisation of costs. Taking these new disciplines as the basis for developing a theoretical underpinning of freight transport models, a whole generation of new freight modelling approaches developed in the late 1980s and early 1990s. Surprisingly, these approaches have hardly made it beyond the scientific literature, due to heavy data requirements or simply a lack of demand by policy makers. In current times, with increasing pressure on government to effectively deal with growing freight flows, the demand is increasing, however, and the available concepts deserve to be more widely known and used.

1.2 Conceptual Framework for Freight Decisions

We will focus on the development of descriptive, empirical models that are built on theories of freight systems behaviour and can be statistically calibrated and validated. The first questions we ask when we go into the behaviour of the actors in the freight system is who the decision-makers are, what decisions they take and how (through which markets) they affect the functioning of the freight transport system.

As we are concerned with long-term effects of transport policy, we also need to focus on changes in the freight transport system on the longer term. The strategic decisions in freight transport involve major investments in assets, such as production plants, are related to longer design or use cycles (e.g. patents and licences) and cannot be reviewed frequently (e.g. because of longer-term contracts or regulations). Such decisions will be made for the longer term, typically once every 5−10 years. The tactical decisions are reviewed on a more frequent basis (typically months to years), as they concern relatively small investments (e.g. a warehouse or a truck), but still have a lag time because the decision is linked to investments or agreements with external parties (e.g. outsourcing contracts, service agreements). Operational decisions are those that can be taken at the discretion of the producing or service providing firm itself and have a short review period in the planning and management cycles (in the order of days to months, like the routing of freight). In line with this distinction between shorter and longer-term decisions, we distinguish three main layers or markets of the freight system (see e.g. other market-oriented frameworks in Manheim (1979), Wandel & Ruijgrok (1993), Liedtke, Tavasszy, &

Wisetjindawat (2009), Roorda, Cavalcante, McCabe, & Kwan (2010), Tavasszy et al. (2012), de Jong et al. (2013) and Savy & Burnham (2013)):

1. Exchange of goods: production, consumption and trade (the commodity market),
2. Inventory networks (the market for inventory logistics services),
3. Choice of transport modes, trips and routes (the market for transport logistics services).

This framework should not be read as the freight version of the four-step passenger transport modelling framework. It is meant as a template for systematic discussion of many decisions, contained within these markets, allowing us to focus on some specific decisions and to integrate others. In addition, the reader should note some substantive differences. Because of the close theoretical relationship between freight generation and spatial distribution, we have combined these into one layer. Also, choice problems related to inventories (the second layer) are specific for freight. Passengers are not stored for later delivery and are (usually) not underway without a destination, whereas freight often is. This phenomenon cannot be tackled by the other decisions.

A short description of these layers, including their demand and supply sides and market mechanisms follows below. We discuss the main decisions, the agents and their importance for freight transport.

1.2.1 Goods: Production, Consumption and Trade

Production and consumption revolve around the actors that form the demand and the supply for goods: producers and consumers. They are both on the sending and the receiving end for freight. Note that producers are on the receiving end of goods flows, such as raw materials and other inputs for manufacturing, and consumers are on the sending end of freight when it concerns waste or return shipments. The decision-makers at firms are the managers responsible for R&D, product, location and plant management etc. The decisions at hand include the location(s) for production, the deployment of factors of production, such as land, goods and services, labour and capital; they will determine the nature of products being made, the volumes of production etc. Consumers and households shape the final demand for goods; their decisions include, by analogy, the residential location, their consumption patterns and the way they deal with waste. An important decision at the level of individual firms and consumers that has to be highlighted here is that of shipment size or order size. We define these two terms in this book as synonyms, so a 'shipment' refers to a bundle of goods that is ordered and delivered together from a producer or warehouse. A transport vehicle or vessel may contain one or multiple shipments (e.g. for several receivers). The total amount of goods in the vehicle is called the 'payload'.

As consumers make trade-offs between the number of times to go the supermarket and the volume of their groceries, so do firms. *Trade* builds on agreements for the transfer of ownership or delivery of service. It determines the spatial boundaries of the flows of goods, and as such drives the spatial organisation of movement of goods. The choice of trading partner and trading volume is also a composite of many underlying decisions. These include the sourcing partners or shopping locations from the demand perspective, from the supply perspective this includes the decisions on sales areas and the price setting for products. The decision-making agents are the managers in industry responsible for sales, marketing or sourcing. They usually belong to the producers, resellers or retailers of the products in question, or in case of goods that involve heavy speculation, professional traders to which this function is outsourced. Consumers and their households are also decision-making agents as they buy products and services. Finally, government also directly influences trade as an agent in the market through import barriers, international taxation rules, customs regulations etc., and by its own production and consumption.

1.2.2 Inventory Networks

Inventory networks are a spatial form of organisation of inventories. Given the spatial patterns and volumes of trade, storage and (de)consolidation of flows may occur at intermediate locations that are in between places of production and consumption. The main purposes of these inventories is to keep logistics costs low by bundling inventories and transport flows and to maintain high service levels with proximity to markets. Whether or not such intermediate inventories are necessary or worthwhile depends on many factors, including the physical characteristics of goods that determine logistics costs structures (e.g. perishability) and the service requirements. The effect that these intermediate inventories have on flows is that the spatial patterns of trade are changed, i.e. new origins and destinations for transport are created. In addition, shipment sizes are determined. These decisions are usually taken by logistics managers of the firms that send or receive goods. Sometimes the role of a logistics manager is partly outsourced to one or more professional service firms (logistics service providers), in which case the decision on locations and volumes of intermediate stocks is taken jointly for several firms.

1.2.3 Transport Organisation

The choice of the modality or mode of transport (road, rail, water, air) is the most discussed point of intervention for freight transport policies. This is in sharp contrast with the industry, however, where the decision is often taken implicitly, or without much contemplation. This can be explained by the fact that, given the dependence on available infrastructures and the transport requirements of the goods, the number of realistic choices for a firm is often limited (Jordans et al., 2006). At the same time, firms are not unwilling to reconsider the mode of transport, as there are substantial differences in scale (and thus costs) and performance between alternative modes. Each mode of transport offers a diverse set of specialised means of transport (= vehicle types within a given mode, e.g. lorries of different sizes), tuned to different good types (e.g. bulk load or unit load) and shipment sizes. As the choice of means of transport is less constrained by infrastructure availability, firms very intensively use this decision to optimise transport. The decision to invest in vehicles is of course a different, longer term one than the assignment of shipments to vehicle or ship types. Note that the choice of mode will often coincide with the decision on shipment size. As with inventories, the agents responsible for logistics management take the decision of mode of transport. This can be outsourced to a logistics service provider or forwarder.

The actual dispatch of shipments, once the mode and means of transport are known, is organised in space and time in the routing and scheduling decision. Transport planners at the shipping company or the carrier (in case of outsourced transport services) carry out this transport planning on a weekly or daily basis. The route followed between intermediate points of loading and unloading may also be suggested by the transport planner to the driver, although the driver will deviate from the route if necessary. There can be multiple origins per destination (n:1, e.g. a retailers distribution centre being served by many producers), multiple destinations per origin (1:n, e.g. a producer who wants to deliver to several clients within a city), or a combination (1:1, e.g. intra-company movements or n:n, e.g. postal services). Different spatial configurations of senders and receivers will require a different trip structure and planning for vehicles. This trip structure can be a simple return trip, but it can also be a complex tour or hub-and-spoke structure. Note that at this stage, a special category of trips is created: empty trips.

The three layers of decision-making discussed above are brought together in a table showing the market at hand, the decisions and decision-makers involved, and the time period of review of decisions.

As will be clear from the above, decisions leading to freight transport are not independent. There can be a direct and mutual dependence between decisions. This is for example the case between production, consumption and trade. While every region has supply and demand of goods, this profile will be different in every region, and trade will take place to take advantage of these differences. According to economic theory, the price of goods in every location will act as a mechanism determining both the local demand and supply, and trade. Modelling these decisions without a connection through price will make the model inconsistent. In the case of transport mode choice and shipment size choice, the optimal shipment size will depend on the costs of transport per shipment, whereas the cost of transport is given by the choice of mode and will depend on shipment size. Ideally, these should be described together. Dependence between decisions also takes place if one decision forms the input for the other; trade determines the volumes between producers and consumers; if these volumes determine the choice of mode or inventory structures, there is a dependency. In the reverse direction, once mode choice and inventory structures are determined, the total logistics costs between origins and destinations are known, which is an input for trade. In short, it is useful to be clear about dependencies between decisions. Does this mean that we have to strive for a completely integrated, comprehensive model? Although theoretically this would be desirable as it would lead to a consistent model, practically this is often unfeasible. In practice therefore, a loose coupling between sub-models of individual markets is attempted. As we will see, however, in practice sub-models are often also integrated in those cases, where an integrative theory is available and where this is feasible in computational or empirical terms.

Note that the literature on these decisions in the management discipline (related to production, logistics and supply chain management) is extensive. Methodologically, this literature focuses on micro-level qualitative analysis (aiming at understanding phenomena through case studies) and on normative (or optimisation) modelling, also at the firm level. There has been little systematic effort to build descriptive, structural models of decision-making, in connection to the knowledge reported in the broader logistics-related management literature. This is the subject of an ongoing stream of research that has grown fast in the past decades. In the subsequent chapters, we will take stock of the state of the art in the field of freight transport models, limiting ourselves to those models that have shown empirical evidence of feasibility and validity.

1.3   Freight Models — Theoretical Perspective of the Book

A compilation of the latest research on most of the freight transport markets and decisions distinguished in Table 1.1 can be found in Ben-Akiva, Meersman, & van de Voorde (2013). This book on the other hand tries to provide an overview of existing freight transport modelling, both traditional and innovative, both well-established and experimental, integrated in a specific overall framework, to be used both by people working at the research frontier and by consultants and researchers doing applied work all over the globe.

The framework provided above of decisions and markets translates as follows into the different model types discussed in this volume (Table 1.2).

At the first level, the alternatives for modelling depend on several assumptions within the system. Assuming fixed prices and technologies will lead to input/output models — if we want to distinguish between sectors. Alternatively, if there is only one sector, the volume and type of economic activity in a region will be the main determinants of the amount of freight entering or leaving a region and we will have freight generation models. Relaxing the assumption of fixed prices and technologies will lead to computable general equilibrium (CGE) models; adding the spatial dimension (i.e. a full interregional framework) will lead to spatial CGE (SCGE). A land use transport interaction (LUTI) model and the regional production function models can be derived from the SCGE model (Bröcker, 1991).

Until here, production, consumption and flows are modelled as a continuous flow, usually measured over a long period of time (typically a year), representing the trade contract. Physical flows often move in discrete quantities, however, and may be stored at discrete points at locations that are central for several producers or consumers. Therefore, at this second level, a conversion is needed to break up the continuous flows into shipments and into the segments of the inventory chains that these shipments will follow. Note that shipment sizes can be infinitely small (resulting in a flow through a pipeline) and that flows also move directly from

Table 1.1 Layers of Decision-Making in Freight Transport — An Outline for Modelling Purposes

Market	Decisions	Decision-Maker	Time Period
Goods: Production, consumption, trade	Plant location Production system Production factors Products Retail outlets Suppliers Shipment sizes to customers	Producer: CEO level, production manager Consumer (household) Producer: marketing, sourcing, sales Consumer (household)	Long term
Inventory networks: Warehousing services	Location of distribution centres, Inventory volumes Assignment of senders and receivers to DCs Intermediate shipment sizes	Marketing manager Logistics manager If outsourced: logistics service provider	Medium term
Transport organisation: Mode, means and route choice	Mode(s) of transport Means of transport (vehicle types within a mode) Intermediate shipment sizes Means of transport Scheduling Routing	Logistics manager Transport manager If outsourced: logistics service provider	Short term

producer to consumer, not just via intermediate inventories. The challenge here is to find out how freight is distributed over these options.

At the third level, when it is clear for individual shipments what their origins and destinations are for transport, transport is organised through the choice of modes, vehicles and routes. We note that these choices will interact with other decisions, such as those concerning the size of the shipment or the choice of distribution centre. We will discuss several ways to deal with these interactions in transport models, e.g. by ignoring the interactions completely or by using fully integrated models.

An important conclusion from this discussion is that many freight models are in fact partial models of a more complex system. Our objective here is to provide a very simple taxonomy with this table that allows to distinguish different models from each other by means of the decision within the system that they represent. In order to obtain a comprehensive view of the whole system, these partial models can be used as building blocks. We will return to the options to create a full freight transport model from the components shown above in Chapter 11.

Table 1.2 Overview of Models Discussed in This Volume

Market	Partial Models	Disciplinary Focus
Production/ consumption and trade	• Production/consumption: Input/ output models • Trade: Gravity models • Combined: Spatial computable general equilibrium models and derivatives • Freight generation models	• Input/output economics • Engineering • Economic geography • Econometrics
Inventory logistics	• Shipment size choice • Inventory chain models	• Operations research • Discrete choice theory
Transport logistics	• Mode choice models • Freight to trip conversion models • Mode and route choice: supernetworks	• Discrete choice theory • Engineering • Network modelling

The right most column in Table 1.2 notes the disciplinary angles, along which the partial models have developed through time. They are worthwhile to note as they still indicate how different professional groups are nowadays working on these models, sometimes in separation. The main disciplines that have contributed to freight demand modelling are transport engineering, growing from the tradition of passenger transport models, and economic geography. Models for production, consumption and trade have traditionally evolved within the domain of economic geography, rooted in microeconomics (Bröcker, 1998; Krugman et al., 1999). For transport engineering purposes, other pragmatic models have been built to be able to forecast trips, or describe feedback relations between freight transport and the economy, in the form of LUTI models. In between the transport engineering modelling and economic geography approaches we find approaches, such as the evolutionary economics models inspired by the techniques of systems dynamics (Fiorello et al., 2010), and the regional production function models (see Wegener, 2011 for an overview).

Aggregate models for trade have been used widely since the 1970s within both domains, in the form of gravity models (Chisholm & O'Sullivan, 1973). It is quite some time ago already, but not widely known, that the transport engineering models have been re-interpreted from the perspective of choice theory and economic geography in aggregate agent models (see Erlander & Stewart, 1990 and Chapter 2, respectively).

The main use (though not exclusive use) of choice theory in freight transport however has been for mode choice. The earliest applications were aggregate models in the engineering domain, later these were replaced by disaggregate models using more sophisticated data acquisition and econometric estimation techniques. Behavioural modelling in freight transport is now entering the field of experimental economics, where the emphasis is on understanding decision rules as an outcome of interactive multiple-agent games and processes, in contrast to econometric estimation of steady states under the assumption of market equilibrium. Finally, another major influence on the state of freight models recently has come from operations research, as it developed normative (optimisation) approaches for firm level decisions on supply chain issues relating to inventories, shipment sizes and transport modes. Note that our perspective here is a descriptive one (our aim being to describe and forecast transport flows for a country, a region or a city). As such we are mainly interested in the question whether normative models for logistics can be used to improve our prediction of decisions of firms under different circumstances.

Although linkages between these disciplines are developing, and economics appears to be an important integrative discipline for freight modelling, we are still far away from a widely accepted, unifying theoretical framework. We will be revisiting the origins of the different model types and note the linkages with other models or disciplines, where appropriate.

Route choice or network assignment was included as one of the decisions in Table 1.1 and as one of the partial models in Table 1.2 (it also is the last stage of the traditional four-stage model). This book does not contain a specific treatment (e.g. in the form of a separate chapter) on traffic assignment to networks. The reason is that most assignment modelling is the same in freight and in passenger transport, but with different restrictions (which can be handled by using the same models, but with different data). For this we therefore refer to a textbook on transport modelling in general, such as Ortuzar & Willumsen (2011). A difference between freight transport assignment and that for passenger transport is that multimodal assignment has received greater attention in freight transport modelling. Some examples of this approach are mentioned in Chapter 6 (on mode choice) of this book.

1.4 Freight Models — Practical Perspectives Addressed

As described earlier, the need for an understanding of the nature of freight transport has emanated from concerns about its growing importance for the economy and the environment.

The needs of policy makers stated above apply in principle to all spatial levels of government: whether at international, state, regional or city level, the concerns and needs for information are very similar. Obviously, depending on the scale-typical policies, governance arrangements and problems, the need for information may be slightly different. We will not pay much attention to these differences, except for one interesting phenomenon: it appears that the smaller and denser the area, the higher the need to integrate the above issues into integrative policy assessment models, with broad stakeholder support. Not surprisingly perhaps, integrative models have been developed mostly from the urban context, combining different methodological angles, including economic geography, transportation engineering and supply chain management approaches into hybrid frameworks (Donnelly, 2007). There are hardly any examples of such comprehensive models outside an urban context; SMILE (Tavasszy et al., 1998) being an exception. For this reason, we devote special attention to urban freight models, and otherwise remain neutral towards the question of tuning models for specific spatial scales.

Government departments responsible for transport policy have increasingly become concerned about the lack of availability of operational tools to forecast freight transport and understand the possible effects of policy measures. Besides understanding the scientific state of the art in freight modelling, there is also a very practical need to develop operational models (be it comprehensive, sketch type or for back-of-the-envelope calculations). In order to cater for this practical demand, we treat a number of implementation issues that modelling practice is confronted with. This concerns the following topics:

1. Approaches to develop simplified freight models,
2. Modelling approaches that allow optimal use of available data,
3. Empirical results on elasticities and service valuation.

Ad 1. The design of the model should follow its purpose, and different purposes require different designs. If one and the same model has to fulfil several purposes, this may lead to unworkable or incomprehensible models. Methods and techniques for reducing model complexity need to be discussed.

Ad 2. Despite the abundance of data in logistics and transport operations, data is often proprietary and difficult to access. We often have to build on publicly available statistics or on large-scale, open surveys. Usually, several databases of different nature and context are available, and data is missing. What data sources are there and which modelling approaches fit best with these sources?

Ad 3. Models can generate numbers for recurrent use in policy making processes; examples are monetary values for converting physical effects of policy into economic effects, or elasticity values that describe the sensitivity of transport flow volumes to changes in transport costs and times.

The discussion on these topics summarises the state of practice and provides examples of concrete applications in freight modelling in various countries.

1.5   Organisation of This Volume

The book is organised according to the framework of the three markets as presented above, and a series of cross-cutting chapters to treat common issues of implementation towards practical usage of models. Each of the Chapters 2−7 relates to one of the markets in the modelling framework. Chapters 8−11 focus on cross-cutting issues related to implementation, summarising the state of practice in this area and showing lines for research and development.

Chapter 2 describes the models used to depict the spatial demand for goods and services: it describes the first layer of our framework. The chapter develops from the inter-sectorial input/output models, through the well-known gravity model for interregional trade, culminating at integrated forms, such as the multiregional input/output (MRIO) and SCGE models used in state-of-the art systems. Chapter 3 takes a different angle at freight and trip production: explaining goods flows without explicitly modelling inter-sectorial relations (which is an important simplification), using freight generation models. If mode choice does not need to be modelled explicitly, trips can be modelled directly with trip generation models, giving a direct relation between economic aggregates and transport movements. For a national freight model, mode choice often is indispensable, but for a model at the urban scale, road transport may be the only mode that really matters. Chapter 4 follows up with a new model type that determines spatial organisation of flows between production and consumption points via inventory networks (the second layer). It explains how freight models can reproduce the spatial patterns of such networks that arise from the need to store goods at intermediate places. Chapters 5 and 6 focus on the mode choice decision (the third layer), where the origins and destinations of freight movements are fixed, except for possible immediate transhipment between modes or cross-docking between carriers. Chapter 5 theoretically and empirically explores the relations between mode choice and the decision of shipment size, which had been ignored in most freight transport models but is moving more to the forefront now. Chapter 6 provides an account of the state of the art in mode choice. Chapter 7 describes the state of the art in modelling the conversion from freight (measured in metric tonnes) to trips (measured in vehicles), including the physical and spatial parameters of these trips (i.e. their shipments and their routing). In addition, the chapter focuses on the freight-typical question of empty trips.

Chapter 8−11 take an integrative stance, compared to the previous chapters. Chapter 8 outlines the development of models for urban freight systems, where the representation of the interests of multiple stakeholders has been an important determinant for model acceptance. Chapter 9 provides an overview of the empirical results of models for use in policy assessment, in the sense of the monetary valuation of freight performance and the elasticity of transport flows for changes in costs. Chapter 10 focuses on the issue of data availability, linking the difficult issue of lack of detailed public data and abundance of proprietary data about freight transport, to alternative solutions in terms of model form. Chapter 11 discusses opportunities for building simplified models where comprehensive models are not feasible and presents lessons learnt from large-scale freight model development, where they are available.

References

Ben-Akiva, M. E., Meersman, H., & van de Voorde, E. (2013). *Freight transport modelling*. Bingley: Emerald.
Bröcker, J. (1991). *Numerische multiregionale Gleichgewichtsanalyse. Lecture notes*. Dresden: Technische Universität Dresden.

Bröcker, J. (1998). Operational spatial computable general equilibrium modeling. *The Annals of Regional Science, 32*, 367–387.

Chisholm, M., & O'Sullivan, P. (1973). *Freight flows and spatial aspects of the British economy*. Cambridge: The University Press.

de Jong, G., Vierth, I., Tavasszy, L. A., & Ben-Akiva, M. (2013). Recent developments in national and international freight transport models. *Transportation, 40*(2), 347–371.

Donnelly, R. (2007). A hybrid microsimulation model of freight flows. In E. Taniguchi, & R. Thompson (Eds.), *City logistics V* (pp. 235–246). Kyoto: Institute of City Logistics.

Erlander, S., & Stewart, N. F. (1990). *The gravity model in transportation analysis*. Utrecht, the Netherlands: VSP.

Fiorello, D., Fermi, F., & Bielanska, D. (2010). The ASTRA model for strategic assessment of transport policies. *System Dynamics Review, 26*(3), 283–290.

Jordans, M., Lammers, B., Tavasszy, L. A., & Ruijgrok, C. J. (2006). *The base potential for inland navigation, rail and short sea shipping (in Dutch)*. Delft: TNO.

Krugman, P., Fujita, M., & Venables, A. (1999). *The spatial economy: cities, regions and international trade*. Boston: Massachusetts Institute of Technology.

Liedtke, G. T., Tavasszy, L. A., & Wisetjindawat, W. (2009). *A comparative analysis of behavior-oriented commodity transport models. Proceedings 88th annual meeting of the Transportation Research Board*. Washington, DC: TRB.

Manheim, M. (1979). *Fundamentals of transportation systems analysis*. Boston: Massachusetts Institute of Technology.

Ortuzar, J. de D., & Willumsen, L. G. (2011). Modelling transport (4th ed.). Chichester: John Wiley & Sons, Ltd.

Roorda, M. J., Cavalcante, R., McCabe, S., & Kwan, H. (2010). A conceptual framework for agent-based modelling of logistics services. *Transportation Research Part E: Logistics and Transportation Review, 46*(1), 18–31.

Savy, M., & Burnham, J. (2013). Freight transport and the moderneconomy. London: Routledge Chapman & Hall.

Tavasszy, L. A., Ruijgrok, C. J., & Davydenko, I. (2012). Incorporating logistics in freight transport demand models: state of the art and research opportunities. *Transport Reviews, 32*(2), 203–219.

Tavasszy, L. A., Smeenk, B., & Ruijgrok, C. J. (1998). A DSS for modelling logistics chains in freight transport systems analysis. *International transactions in operational research, 5*(6), 447–459.

Wandel, S., & Ruijgrok, C. (1993). Innovation and structural changes in logistics: a theoretical framework. In G Giannopoulos, & A. Gillespie (Eds.), *Transport and communications innovation in Europe* (pp. 233–258). London: Belhaven Press.

Wegener, M. (2011). Transport in spatial models of economic development. In A. de Palma, R. Lindsey, E. Quinet, & R. Vickerman (Eds.), *Handbook in transport economics* (pp. 46–66). Cheltenham: Edward Elgar.

2 Modelling Inter-Regional Freight Demand with Input−Output, Gravity and SCGE Methodologies

Olga Ivanova

Dutch Organization for Applied Scientific Research (TNO), Strategy and Policy, The Netherlands

2.1 Introduction

Continuing population and economic growth, trade specialisation and globalisation trends of the last decades have resulted in an unprecedented increase in freight transport movements both in terms of volume as well as in terms of average distance. Reductions in costs and improvements in the efficiency of freight transportation have acted as drivers of further globalisation, economic growth and contributed to the global integration of the capital market and an increase in the amount and scope of foreign direct investments (FDIs) in different parts of the world.

International and inter-regional freight demand is determined by the spatial distribution of production and consumption activities. International freight flows mirror to a large extent the trade flows between the countries with the exception of a relatively small number of transit countries with large ports, such as Singapore, Panama and the Netherlands, where some goods are being transhipped. The main driving forces of freight transport demand include population growth and population migration, technological development, specialisation including access to natural resources, agglomeration and dispersion forces. These drivers are quite similar to the drivers of international trade which determines the methodologies that are currently being used for prediction of international and inter-regional demand for freight transportation services.

It is well established in the literature that forecasting freight transport flows is more difficult from a methodological point of view than modelling passenger transport demand. This complexity is partially explained by the large number of actors that are involved in the generation of freight transport flows as well as the large number of heterogeneous goods that are being transported by a combination of different transport modes. The complexity of the freight movements is one of the main reasons why the development of freight transport modelling has been lagging behind the development of passenger transport modelling.

Existing methodologies also differ with respect to the aim of the overall analysis. Most of the existing freight transport models are designed to assess the effects

Modelling Freight Transport. DOI: http://dx.doi.org/10.1016/B978-0-12-410400-6.00002-1

of what-if scenarios, such as for example changes in the transport costs, rather than to provide a long-term forecast of the developments in freight demand. The later task is more difficult and requires knowledge on the long-term changes in the worldwide production and specialisation patterns. Usually this is done via a construction of long-term scenarios outside of the modelling tool. The realisation of a specific scenario is uncertain and their complexity varies depending on the type of the policy or research question at hand.

In 2008, Paul Krugman was awarded the Nobel Prize in economics for his work on the 'analysis of trade patterns and location of economic activity'. The development of his groundbreaking theory and related models had a major influence on our understanding of the role that freight transport costs play in economic development and location decisions of firms and individuals. The work of Krugman laid the foundation for the new economic geography (NEG) theory which gives an explanation for the mechanisms of agglomeration and dispersion forces behind the geographical patterns of various types of production activity. Since distance and associated transport costs play a central role in the explanation of agglomeration patterns NEG theory is currently used in several models that predict changes in international and inter-regional freight demand flows.

The spatial patterns of production and consumption activities give rise to international and inter-regional trade and freight transportation flows. The general methodology of modelling freight demand involves the following two steps: (1) representation of spatial distribution of economic activities that is spatial demand and supply of commodities and their changes as response to policies and macroeconomic trends; (2) given the spatial distribution of demand and supply for various commodities the trade flows between countries and/or regions are calculated on the basis of models of micro or representative economic behaviour of the trader.

The main approaches that are used for modelling of trade flows between regions and hence freight demand given a particular spatial distribution of supply and demand include:

- structural spatial models and in particular gravity model;
- discrete choice models, such as logit and nested logit;
- (nested) constant elasticity of substitution (CES) functions used as a part of spatial computable general equilibrium (SCGE) models.

Spatial interactions model is an aggregated one and includes the representation of aggregated spatial zones and economic activities. The most popular spatial interaction model is the gravity model that assumes that the trade flows (or other integration flows such as migration for example) between two zones depend positively on the level of economic activity in these zones and negatively upon the distance (proxy for transportation costs) between them. Gravity models have been widely used in the last decades in order to forecast the development and changes as result of policies of the international trade flows.

Discrete choice models describe, explain and predict choices between two or more discrete alternatives, such as choosing between modes of transport and/or origin of a particular final/intermediate commodity. Daniel McFadden has won the

Nobel Prize in economics in 2000 for his work in developing the theoretical basis for discrete choice models. These models including logit and nested logit types relate the choice made by each person to the attributes of the person and the attributes of the alternatives. For example, the choice of the wholesaler from which region to purchase commodities depends on the regional prices, transportation costs between the regions and some unobserved characteristics of the regions. Logit and nested logit models estimate the probability that a person chooses a particular alternative.

The CES production function was introduced by Arrow, Chenery, Minhas and Solow in 1961. Formally, the elasticity of substitution measures the percentage change in factor proportions due to a percentage change in the marginal rate of technical substitution. This function has nice mathematical properties and has been widely used in general equilibrium models. CES function has also been the core of the NEG models and allowed one to capture the representation of the complex agglomeration forces. It has been demonstrated by Anderson, de Palma, & Thisse (1987) that the CES demand function is a special case of a nested logit model whose second stage is deterministic.

The main methods that are currently used in order to investigate the spatial distribution of economic activity and inter-regional trade flows include:

- microeconomic theory in particular the Von Thünen and Alonso models,
- spatial accounting models in particular inter-regional input–output (IO) and SCGE models.

Microeconomic theory focuses on the representation of the individual behaviour of consumers and producers. The applications of this theory to the location of activities includes Von Thünen model which captures the impact of transportation costs on land use and land prices. Alonso has further introduced the bid-price curve for the land prices that has been widely used in the land-use transport-interaction models, such as MEPLAN and URBANSIM (Sivakumar, 2007). The models based on the microeconomic theory include the decisions of households and firms that are based on utility and profit maximisation assumptions in combination with equilibrium on the land market that results in the equilibrium land prices.

The spatial accounting models include inter-regional IO models and SCGE models. IO models describe the inter-industry relationships in terms of intermediate inputs. The IO model has been originally proposed by the Nobel Prize winner Leontief and has been focuses on the single country interactions between the industries. The output and intermediate inputs of the industries in his analytical framework are driven by the developments in final demand of the households.

Standard multi-regional IO models are not flexible as their inter-regional trade coefficients are constant and do not reflect the changes in regional prices and inter-regional trade and transport costs. Random utility can be adopted in order to describe how sectors choose in which region to purchase their intermediate inputs in a utility maximising or cost minimising way. This can be done using the discrete choice models including logit and nested logit. These models are estimated econometrically and integrated within the general structure of the IO model.

The background of SCGE models can be found in the theory of Walras. Its core is the concept of market equilibrium where the 'invisible hand' of Adam Smith acts an auctioneer and keeps demand equal to supply on all the markets in the economy. The SCGE model belongs to the class of microeconomic-based macro-models since it includes the representation of microeconomic behaviour of economic agents, such as utility maximisation and profit maximisation as well as the representation of all markets in the economy.

The rest of the chapter is organised as follows. We start with an overview of the present state-of-the-art in international and inter-regional freight demand modelling. This overview pays particular attention to the differences between the theoretical and empirical literature as well as the pros and cons of each of the existing methodologies. We further focus on the three main approaches to the modelling of freight demand including gravity model, the IO framework and spatial general equilibrium modelling. Each of the methodologies is explained in detail and the implementation challenges are discussed. In the concluding part of the chapter we provide the comparison between the three main methodologies and present our view of the future research agenda in the field of gravity and spatial general equilibrium modelling.

2.2 State-of-the-Art in of Inter-Regional Freight Demand Modelling

During the past decades the modelling of trade flows and consequently freight flows has been widely researched and different methodologies have been developed in order both to forecast the future trade flows between the countries as well as to analyse the changes in trade patterns as a result of some macroeconomic and policy changes. The forecasting of the trade flows is done with the use of econometric models, such as spatial interaction models and in particular the gravity model. The parameters of these models are based on the historical data and are usually statistically significant.

The 'what-if' scenarios are usually analysed with the use of the simulation models including inter-regional IO models and SCGE models. Simulation models focus on capturing the structure and main drivers of the trade flows including the location of various production and consumption activities, the level of transportation and trade costs as well as the impact of various policy instruments, such as taxes and subsidies. The advantage of using SCGE models for the calculation of the trade flows is that they allow for endogenous calculation of impacts of transport costs and trade tariffs on the regional price levels, production and income distribution between the regions.

One of the widespread critiques of the use of SCGE models for predicting changes in trade flows is that most of their parameters are not estimated econometrically and hence their results cannot be validated statistically. Despite this critique the result of applied SCGE models have been found robust and theoretically correct in many trade-related studies. Besides that, the possibility exists (1) to use econometrically estimates parameters as a part of SCGE models and (2) to integrate econometrically estimated models of trade within the SCGE structure.

Econometric approaches to modelling international trade (Erlander & Stewart, 1990) have mostly focused on the gravity equation and prediction of the total trade flow between the pair of countries as a function of their Gross Domestic Products (GDPs), bilateral distance (as a proxy for transport costs) and some other explanatory variables, such as non-monetary trade barriers (custom procedures), history, culture and institutions.

The importance of transport infrastructure for productivity and economic growth has been quite widely investigated in both theoretical and empirical literature (Banister & Berechman, 2001). It has been pinpointed as one of the main factors behind the production specialisation and the volume of international trade. The main mechanism behind this finding is that the stock of transport infrastructure is the main explanatory factor for the level of transport costs.

In the majority of trade literature transport costs have been considered as exogenous variables and approximated by the geographical distance between the geographical locations. This assumption has been largely justified by the absence of reliable data on the transportation costs as well as the focus of the present trade-related research on trade factors other than transport infrastructure and logistics.

Most of the existing empirical literature on the gravity model focuses on aggregated trade flows data and ignores the differences between various commodities and region pairs in terms of the impact of transport costs on trade patterns and volumes. This aggregated approach does not allow for making policy-relevant conclusions related to particular industries and commodities and diminishes the usefulness of the econometric analysis.

The focus of researchers on the aggregated trade flows might be explained by the large heterogeneity in the results of econometric estimates on the detailed trade data by commodity type in combination with a large share of non-significant parameters in case of regressions for detailed commodity groups. Another challenge is the presence of a significant number of zero trade flows in the detailed data which requires the use of more sophisticated econometric estimation techniques. Instead of using ordinary least squares (OLS) for the estimation a researcher should then make use of the more advanced Poisson, Tobit or Heckman's two-stage estimation models (Helpman, Melitz & Rubinstein, 2008; Santos Silva & Tenreyro, 2006) in order to properly estimate the gravity model in the presence of zero trade flows.

Another strand of literature is related to the use of multi-regional IO model with fixed and variable trade coefficients. The basic concept of this modelling framework can be traced back to the theory of Keynes that has introduced the concept of effective demand postulating that production is determined by consumption. Leontief has translated this theory to the multi-sectoral setting and proposed the country-level IO framework which was able to simulate the inter-industry relationship via the use of fixed technical input coefficients. The work on the multi-regional IO models was pioneered by Chenery, Isard and Moses in the 1950s. These works have introduced inter-regional trade coefficients that allow for calculation of inter-regional trade but did not specify any functional forms behind them.

The development of the random utility theory (see Chapter 6) has allowed for development of the functional forms for the trade coefficient that are based on microeconomic behavioural principles. In the new generation of multi-regional IO models with variable trade coefficients, inter-regional trade flows respond to the changes in inter-regional trade and transport costs.

Even though the multi-regional IO models with variable trade coefficients took into account the impact of transport costs on trade patterns and volumes, they still did not account for changes in prices, links between changes in households' incomes and expenditures and were purely demand-driven. These drawbacks gave rise to another strand of modelling literature, namely SCGE modelling. The background of these models can be found in the theory of Walras. Its core is the concept of market equilibrium where the 'invisible hand' of Adam Smith acts as an auctioneer and keeps demand equal to supply on all the markets in the economy. The SCGE model belongs to the class of microeconomic-based macro-models since it includes the representation of microeconomic behaviour of economic agents, such as utility maximisation and profit maximisation as well as the representation of all markets in the economy.

Computable general equilibrium (CGE) models offer a framework to model the full level of economic interactions, based on IO tables and national account data. SCGE models typically are comparative static equilibrium models of inter-regional trade and location based in microeconomics, using utility and production functions with substitution between inputs, mostly modelled using CES equations. Since the 'revival' of regional economics, particularly attributed to the work of Fujita, Krugman & Venables (2001), many SCGE models incorporate the monopolistic competition framework of Dixit–Stiglitz to represent the effects of regional competition and concentration of economic activity.

During the past decade, several SCGE models have been developed for the analysis of policy-related questions. Some examples of well-known CGE models with disaggregation on the level of regions are CGEurope (Bröcker, Korzhenevych & Schuermann, 2010), RAEM (Ivanova et al., 2007), GEM-E-3 (Capros et al., 1997), etc. The present SCGE models have a sophisticated theoretical foundation and rather complex, non-linear mathematics. The latter is precisely the reason why SCGE models are able to model (dis)economies of scale, external economies of spatial clusters of activity, continuous substitution between capital, labour, energy and material inputs in the case of firms, and between different consumption goods in the case of households.

The main basis and logical assumptions of SCGE can be summarised in a few lines:

1. Interaction between regions is costly and requires transport, which is essential in the model.
2. The further two regions are apart, the more costly the interaction (increased transport and trade costs).
3. Trade interactions are modelled with CES functions, applying the Armington assumption. The Armington assumption states that imported and domestically produced products are imperfect substitutes (or varieties) and that consumers prefer a 'balanced' amount of these

goods. The later means that the consumer cannot fully replace domestic goods with the imported ones and vice versa.

4. Passenger and freight transport behave differently. Freight transport is often modelled in much more detail and is the main source of interaction between regions.
5. Supply should equal demand at the regional level, reflecting market clearance.
6. In the framework of Dixit–Stiglitz monopolistic competition, increased concentration of industries increases the variety to consumers and leads to bonuses in terms of proximity.

The applications of SCGE models are mainly focused on regional infrastructure investments, transport policy-related issues, such as road taxes or road charging and regional planning (housing, labour market policy and land use).

2.3   Forecasting Inter-Regional Trade Using Gravity

2.3.1   The History of Gravity Model

The gravity model has been first introduced into economics by the Nobel Prize winner Tinbergen and Linneman to explain the flows of international trade in goods. Despite the lack of a solid foundation in behavioural theory, the gravity model has proved to be quite robust and empirically sound. During the last decades the gravity model has been applied in numerous studies in order to explain the factors that influence the changes in trade flows which include in particular the distance and related transport costs. The gravity model has been also implemented for the modelling of trade in services, migration flows and capital (in particular FDIs) flows between the different countries of the world.

The gravity model is based on the equations from the Newtonian physics. His law of universal gravitation describes the gravitation force between the two masses in relation to the distance that lies between them the following way:

$$F_{ij} = G \frac{M_i M_j}{d_{ij}^2} \tag{2.1}$$

The gravitational force F_{ij} between the two bodies is a function of their masses M_i and M_j divided by the square of the distance between them d_{ij} and multiplied with the gravitational constant G that is identified empirically.

In case of trade flows this is translated into an equation where the bilateral trade flow is a function of the sizes of the trade partners (their GDPs) and their proximity (distance). The basic gravity equation can be formulated as follows:

$$F_{ij} = O_i \cdot D_j \cdot R_{ij} \tag{2.2}$$

where F_{ij} is a trade flow of goods, O_i is the vector of the characteristics of the origin and D_j is respectively the vector of the characteristics of the destination, R_{ij} is the bilateral resistance factor that can itself be a function of various explanatory variables, such as distance, transportation and trade costs, non-monetary barriers to trade, etc.

In order to simplify the econometric estimation of the gravity model, the equation above can be log-linearised in the following way:

$$F_{ij}^* = \log F_{ij} = \alpha + \beta_1 \cdot \log O_i + \beta_2 \cdot \log D_j + \beta_3 \cdot \log R_{ij} + \varepsilon, \varepsilon \in N(o, \sigma^2) \quad (2.3)$$

2.3.2 Theoretical Foundations of Gravity Equation

The theoretical foundations of the gravity equation are based on the reduced form of the spatial general equilibrium model with iceberg transport costs, monopolistic competition and increasing returns to scale. Let us follow the modelling set-up followed by Bergstrand et al. (2013).

Let us assume that there exists a single regional consumer with the CES utility function depending on the amount of goods produced in different regions of the economy. There are iceberg transport costs associated with movement of goods between the symmetric regions. This means that a certain share of good disappears during the movement to reflect transport costs. Regions and firms in the model are symmetric meaning that all products in region i sell at the same price p_i. The value of the total trade flow between region and j denoted X_{ij} is equal to $n_i p_i t_{ij} c_{ij}$, where n_i is the number of firms/varieties produced in region i, $t_{ij} \geq 1$ are *ad valorem* iceberg trade costs (i.e. as a percentage of the value of trade flow) and c_{ij} is the demand in region j of the output if each firm/variety in the region i. Given the CES utility function in each of the region, the trade flow can be derived as a function of regional effective demand D_j that is equal to the GDP of the region:

$$X_{ij} = n_i \left(\frac{p_i t_{ij}}{P_j} \right)^{1-\sigma} D_j \quad (2.4)$$

where $P_j = \left[\sum_{n=1}^{N} n_i (p_i t_{ij})^{1-\sigma} \right]^{1/1-\sigma}$ is the CES composite price index derived on the basis of the consumer's utility function.

It is assumed that each product/variety is produced under the increasing returns to scale in a monopolistically competitive market using only one production factor, labour. This assumption allows one to identify the equilibrium number of firms/varieties on the market, n_i.

The representative firm in region i is maximising its profits under the linear costs function:

$$l_i = \alpha + \phi y_i \quad (2.5)$$

where l_i denotes the amount of labour used for production by the representative firm in region i, α is the amount of labour necessary to set up a new firm, y_i is the output of the representative firm in the region i and ϕ is the labour costs per unit of output. It is assumed that both fixed and variable labour costs are homogenous across the regions.

The profit maximisation of the representative firm results in the condition for the optimum prices of the output that are determined as the monopolistic markup over the unit production costs:

$$p_i = \frac{\sigma}{\sigma - 1} \phi w_i \tag{2.6}$$

where w_i is the wage in region i and ϕw_i represent the costs of a unit of production output in the region.

Under the monopolistic competition zero economic profits in the equilibrium with free entry of firms ensure that

$$y_i = \frac{\alpha}{\phi}(\sigma - 1) = y \tag{2.7}$$

and the output of each monopolistic firm is the same and is equal to y. An assumption about the full utilisation of the labour endowment in each of the regions L_i results in the following equilibrium number of firms/varieties:

$$n_i = \frac{L_i}{\alpha + \phi y} = \frac{L_i}{\alpha \sigma} \tag{2.8}$$

Based on the assumptions and derivations presented above it is possible to rewrite the formula for the total trade flows between the regions i and j as the function of regional-realised demands/GDPs, labour endowments and trade costs:

$$X_{ij} = D_i D_j \frac{(D_i/L_i)^{-\sigma} t_{ij}^{1-\sigma}}{\sum\limits_{k=1}^{N} D_k (D_k/L_k)^{-\sigma} t_{kj}^{1-\sigma}} \tag{2.9}$$

The formula above represents the gravity equation that can be used for empirical analysis.

2.4   Multi-Regional I/O Framework

2.4.1   Multi-Regional IO Models with Fixed Trade Coefficients

IO models represent the economic interactions between different economic agents, such as firms and consumers in the economy. The models employ the representative actor assumption where the behaviour of heterogeneous firms in each sector can be averaged out and represented as behaviour of one large firm. Assume that we have M production sectors and one representative consumer in the single region economy. The IO model describes the monetary flows of commodities and services between the economic agents in the following way:

$$X_m = \sum_n x_{mn} + Y_m \tag{2.10}$$

where X_m is the total output of sector m that is defined as the sum of intermediate uses of its products by all sectors in the economy x_{mn} plus the final consumption of the households, Y_m. The direct purchases of the sectors can be expressed via the use of Leontief technical coefficients a_{mn} as

$$x_{mn} = a_{mn}X_n \tag{2.11}$$

The initial equation can be than rewritten in a classical way as follows:

$$X_m = \sum_n a_{mn}X_n + Y_m \tag{2.12}$$

The extension of IO model to multiple regions was first proposed by Isard in 1960, who has introduced an explicit spatial dimension into the inter-sectoral flows table in the following way:

$$X_{im} = \sum_j x_{mij} \tag{2.13}$$

where x_{mij} is the trade flows of production of sector m from region i to region j. This can be rewritten in the classical way with the use of the technical coefficients a_{mnj}:

$$x_{mij} = \sum_n a_{mnj} \sum_k x_{njk} + Y_{mj} \tag{2.14}$$

If one denotes the total consumption of commodity m in region j as $C_{mj} = \sum_i x_{mji}$, this can be rewritten as:

$$C_{mj} = \sum_n a_{mnj}X_{nj} + Y_{mj} \tag{2.15}$$

2.4.2 Introduction of Variable Trade Coefficients Using Random Utility Theory

Standard multi-regional IO models are not flexible as their inter-regional trade coefficients are constant and do not reflect the changes in regional prices and inter-regional trade and transport costs. In order to overcome this important draw-back of the inter-regional IO models several empirical applications have made use of the random utility theory in order to capture the impacts of changes in transport costs on the share of commodities bought by firms and households in region i from the rest of the regions in the model.

Random utility can be adopted in order to describe how sectors choose in which region to purchase their intermediate inputs in a utility maximising or cost minimising way. Based on discrete choice theory the trade volume of sector n from region i to region j is written as:

$$x_{nij} = C_{nj} \frac{\exp(\lambda_n v_{nij})}{\sum_k \exp(\lambda_n v_{nkj})} \qquad (2.16)$$

where λ_n is a dispersion parameter and v_{nij} is the systematic utility that includes the disutility of transport costs between the regions. It should be noted that the derivation of structural gravity model can be also based on modelling of individual discrete choices of the trader that faces unobservable costs and benefits presented above (Erlander & Stewart, 1990). The trader chooses the bilateral trade pair with the highest benefit. The discrete choice model is used in order to represent the behaviour of the population of individual traders and can be estimated econometrically. Building on the multinomial logit one can derive a structural gravity model based on the efficiency principle.

Random-utility multi-regional IO models still suffer from a number of important deficiencies, in particular:

- They do not sufficiently take into account the interdependencies between incomes and expenditures.
- They are demand-driven models and hence the supply side effects cannot be modelled properly.

In the recent decade significant progress has been made in the field of computable SCGE modelling that can be considered as a good alternative to inter-regional IO models since they preserve the modelling capacities of inter-regional IO models and compensate for their drawbacks mentioned above.

2.4.3   Overview of Empirical Inter-Regional IO Models

2.4.3.1   RIMS II

RIMS II is the regional IO modelling system for the United States (U.S. Department of Commerce, 1997) that is used for the calculation of different type of multipliers related to the regional effects of public and private sector projects and development programs. RIMS II is based on an IO table derived from two data sources: BEA's national I—O table, which shows the input and output structure of nearly 500 US industries, and the BEA's regional economic accounts, which are used to adjust the national I−O table to show a region's industrial structure and trading patterns. One of the examples of use of RIMS II is the evaluation of regional impacts of military base closings, impacts of expanding airports and effects of the investments into shopping malls and sports stadiums.

2.4.3.2 IMPLAN

IMPLAN is a regional IO model for the United States.[1] IMPLAN uses a uniform national production technology and applies the data on regional coefficient approach in order to regionalise the IO coefficients for the production sectors. IMPLAN uses a top-down approach to build up its database. National data are used as a restriction for the state data. Employment and earnings data are based on County Business Patterns data and BEA data. IMPLAN estimates output at the state level by using value-added report by BEA as proxies to allocate US total gross output. Also, IMPLAN allocates state total gross output to counties based on county employment earnings.

2.4.3.3 REMI Model

The REMI model was developed in 1980s for commercial applications and has been extensively marketed to a large number of customers by the same named firm.[2] As a US regional model, the main market was the US state governments, especially the state departments of transportation, but also a handful of European customers purchased the services of REMI. Despite its commercial success and continuous revision, conceptually the model is based on the state of the regional science of the 1970s and 1980s.

REMI is a hybrid model that links an IO core to a regional econometric model. Without the econometric responses REMI collapses to a standard IO model. The econometric specifications are neoclassical in nature. The model uses the notion of regional for the long-term representation of regional economic growth.

2.4.3.4 FIDELIO 1

Fully inter-regional dynamic econometric long-term input-output (FIDELIO) model for the EU27 is a new hybrid econometric and IO model developed for Joint Research Centre (JRC) − Institute for Prospective technological Studies (IPTS) that uses the latest World Input Output Database (WIOD) data set (www.wiod.org) for the estimation of its econometric core (Kratena et al., 2013). It uses a set of supply-use tables at the level of 59 NACE industries/commodities for the EU27. For the representation of factor demand and trade FIDELIO uses flexible functional forms, such as Translog and AIDS.

Private consumption in the model is a result of a dynamic optimisation process with durables and non-durables. FIDELIO focuses on energy and the energy efficiency of the durables purchases by households. The production functions in the

---

[1] See http://implan.com/V4/Index.php Accessed 24.07.13.
[2] See Regional Economic Models Inc. http://www.remi.com/ Accessed 24.07.13.

model are described by a Translog model that differentiates capital, labour, energy, imported and domestic intermediates. That gives a different model of international trade as compared to the standard CGE model that makes use of the CES functions. The model includes satellite accounts for energy and environment based on WIOD data set. Labour supply is modelled under the assumption of wage bargaining based on the wage curve model.

2.4.3.5   IDEM

Integrated demographic and economic model (IDEM) of Italy is a multi-regional model of the Italian economy that consists of the demographic part that is based on spatial cohort-component approach, and an economic part that is based on a multi-regional IO approach (Fachin & Venanzoni, 2012). The two parts of the model are interlinked by productivity growth, labour market participation and migration flow variables. The model has been used for regionalisation of national economic forecast of the Italian Treasury and evaluation of regional public investments and interventions, such as, for example, investments into transport infrastructure.

2.4.3.6   PECAS

The production exchange and consumption allocation system (PECAS) model for Canada and United States has been developed by Dr. Doug Hunt and Dr. John Abraham of the University of Calgary (Hunt & Abraham, 2009). It consists of two principal models: (1) activity allocation (AA) model: an aggregate, equilibrium spatial IO model and (2) spatial development (SD) model: a disaggregate state-transition model. The core of PECAS is a spatial IO model that represents the monetary trade flows between different regions. The monetary inter-regional trade flows are further translated into the commodity flows in physical units. The model is static and calculates regional consumption prices on the annual basis. PECAS includes the representation of labour market and real estate market as well as the markets for other goods and services.

2.5   Spatial General Equilibrium Models and NEG Effects

2.5.1   Overview of the Methodology

SCGE is a regionalised version of the CGE model and has all its major features. SCGE models are usually calibrated on the multi-regional social accounting matrix (SAM) for 1 year. (S)CGE models incorporate microeconomic behaviour of all major agents in the economy as well as the main macroeconomic relationships, such as market balance, the relationships between savings and investments, trade balance and interest rate, governmental balance, etc. During the last decades they have proved to be an interesting and important tool for policy analysis.

    According to Thissen (1998, p. 2) a CGE model is 'the general equilibrium links among income of various groups, the pattern of demand, the balance of payments and a multi-sector production structure. Moreover, the model incorporates a set of behavioural equations describing the economic behaviour of the agents identified in the model and the technological and institutional constraints with which they are faced. The model is in general equilibrium, because a set of prices and quantities exists, such that all excess demands for commodities and services, in nominal as well as in real quantities, are zero.'

CGE models consist of the behavioural investment, demand and supply equations, where the coefficients are either estimated econometrically or calibrated (the calibration technique is explained later in this section). Fundamental macroeconomic behavioural equations are based on the microeconomic theory of producer and consumer. CGE models incorporate linkages between the production sectors in the form of inter-sectoral trade. They describe how much of each commodity and service a sector needs to buy both per unit of the output and in total. The amounts of commodities and services used per unit of production output are determined endogenously and depend upon the production technology of the sector and relative prices of commodities and services.

The choice of the closure rule of the (S)CGE model reflects the macroeconomic theory the model-builder believes in. For example, according to the neoclassical theory labour is paid a wage equal to its marginal productivity and is flexible, whereas according to the neo-Keynesian theory nominal wage is fixed and does not move. The choice of the macroeconomic closure affects the possibilities for policy simulations with the (S)CGE.

The macroeconomic part of the (S)CGE model includes the relationship describing the development of the capital stock, where the stock at year t is equal to the stock in year $t - 1$ minus depreciation plus investment. It also includes the saving decisions of the households and firms. In the simplest case, firms save enough to cover their depreciation plus expected economic growth rate and households save a constant share of their disposable income each period of time. Total savings are spread between different sectors (and regions) based on the expected rate of return and depreciation in a particular sector (and region). The rule according to which savings are allocated to different sectors varies between the (S)CGE models and reflects the assumptions of the model-builder about expectations and behaviour of the economic agents.

2.5.2   Calibration of SCGE Models

Calibration is the most widely used methodology to derive a consistent set of parameters of the SCGE model. In order to derive these parameters one has to assume that the initial data set of the model represents the equilibrium point and hence can be viewed as a reference equilibrium. The parameters of the model are calculated in such a way that the model solution reproduces the empirical data set of the model. The core of an SCGE model database is the SAM that represents the monetary flows in the economy and makes sure that a number of equilibrium conditions are satisfied. The columns of the SAM represent the economic sectors whereas the rows represent the markets for goods, services and factors of production. The columns of the SAM provide information about the intermediate and factor inputs. The total sum of production inputs equals the sum of sector-specific outputs which corresponds to the equilibrium zero profit condition. The rows of the SAM provide information about supply of and demand for goods, services and factors of production. The total sum of supplies is equal to the total sum of demands which corresponds to the market equilibrium condition.

The calibration procedure has generated a lot of criticism on the literature due to the impossibility of performing a sound statistical test on the values of the model parameters. This requires modellers to make sure that each model simulation is accompanied by thorough sensitivity analysis. An alternative technique is an econometric estimation of model parameters based on time series, cross section and/or panel data. This technique is widely used for macroeconomic general equilibrium models with small number of equations but impossible to use in the case of SCGE models due to (1) lack of data necessary for full econometric estimation and (2) lack of powerful solution algorithms for the estimation of large systems of equations. As a compromise some existing empirical SCGE models use a number of econometrically estimated parameters and calibrate the rest of them on the initial data of the SAM.

2.5.3 Simple Mathematical Formulation

The following mathematical formulation of a simple SCGE model is an extension of the model used by Krugman for the derivation of the gravity equation and follows the notations and structure explained in Section 2.3 of this chapter. It illustrates the main elements of the modelling framework which should be considered as the core of the model and by no means as the ultimate structure. In real applications the model structure reflects the research question at hand and the data availability of each particular country.

The core SCGE model represents a closed economy with N sectors, $n = 1, 2, \ldots, N$ and I regions, $i = 1, 2, \ldots, I$. Each sector is assumed to produce only one type of commodity hence there is a full correspondence between sectors and commodities. In each region there are three types of the activities: production, transport and final consumption. The inter-regional part of the model is based on the so-called pooling concept. This concept states that the commodities of type i used for intermediate and final use in region r are first merged into a pool of commodity i in region r and only after that delivered to intermediate and final consumers. This assumption implies that the regional mix of each commodity does not vary between the final and intermediate consumers in a particular region. The pooling concept greatly simplifies the data requirements and the dimensionality of SCGE model.

Each region i of the model includes N production sectors, one household and N transport agents or wholesalers who choose the regional mix of each commodity consumed in the region r. Sector i in region r produces its output using pooled goods and labour as its production inputs. The regional household earns its income from selling its labour endowment and purchases final consumption goods from the region-specific pool.

Each regional transport sector or wholesaler operates under the CES utility function depending on the amount of goods produced in different regions of the economy. There is commodity-specific *ad valorem* iceberg transport costs t_{nij} associated with movement of goods between the symmetric regions. This means that a certain share of the goods disappears during the movement. Given the CES utility function of the transport sector or wholesaler in each of the regions, the trade flow can be derived as:

$$X_{nij} = \left(\frac{p_{ni}t_{nij}}{q_{nj}}\right)^{1-\sigma_n} S_{nj} \tag{2.17}$$

where, $q_{nj} = \left[\sum_{n=1}^{N}(p_{ni}t_{nij})^{1-\sigma_n}\right]^{1/1-\sigma_n}$ is the CES composite price index derived on the basis of the CES utility function of the transport sector or wholesaler and S_{nj} is the total sales of the commodity n in the region j. The total regional sales consists of intermediate consumption by the production sectors Q_{nkj} plus the final consumption by the regional household C_{nj} that is $S_{nj} = \sum_k Q_{nkj} + C_{nj}$. The CES composite price index q_{ni} represents the equilibrium prices of the pooled commodities in region i.

The output of commodity n in region iY_{ni} is produced according to the Cobb−Douglas production function that uses labour L_{ni} and the aggregate of intermediate inputs Q_{nkj} as its factors:

$$Y_{ni} = L_{ni}^{\alpha_n}\left(\prod_k Q_{kni}^{\gamma_{kn}}\right)^{1-\alpha_n} \tag{2.18}$$

where α_n and γ_{in} are specific Cobb−Douglas shares of labour and intermediate inputs.

Optimal intermediate and labour inputs of the firms are derived as a result of costs minimisation under the technology constraint represented by the production function in the following way:

$$L_{ni} = \frac{\alpha_n Y_{ni} p_{ni}}{w_i} \tag{2.19}$$

$$Q_{kni} = \frac{(1-\alpha_n)\gamma_{in} Y_{ni} p_{ni}}{q_{ki}} \tag{2.20}$$

Based on the functional form of the production function, the unit production costs are calculated as:

$$c_{ni} = w_i^{\alpha_n}\left(\prod_k q_{ki}^{\gamma_{kn}}\right)^{1-\alpha_n} \tag{2.21}$$

Regional households use the income that they get from selling their labour endowment $\overline{L}_i$ to the firms in their region on purchasing pooled commodities from the transport sector or wholesaler. The amount of purchased good is derived on the basis of the following utility maximisation problem:

$$U_i = \prod_k C_{ki}^{\beta_k} \rightarrow \max, \sum_k C_{ki}q_{ki} = \overline{L}_i w_i \tag{2.22}$$

The optimal households' demands are derived as:

$$C_{ki} = \frac{\beta_k \overline{L}_i w_i}{q_{ki}} \tag{2.23}$$

Equilibrium market conditions of the model include equilibrium on the regional labour market which determines the level of the regional wages w_i:

$$\sum_n L_{ni} = \overline{L}_i \tag{2.24}$$

And equilibrium on the commodity market that determines the level of the producer prices p_{ni}:

$$Y_{ni} = \sum_k X_{nik} \tag{2.25}$$

The latter equilibrium condition states that the level of the output in a particular region is equal to the sum of all outgoing trade flows plus the consumption in the own region. These trade flows represent the region-specific demands for the output produced in the region. The combination of intermediate and final demand equations with the market equilibrium conditions gives the formulation of a simple SCGE model.

2.5.4 Including the Spatial Dimension

A geographic or spatial dimension has been included in formal economic analysis in a number of different ways over the years. Many papers have dealt with geography and space or distance through proxies, like transport costs or land price gradients or labour mobility. A geographic pattern of economic activity arises through a tension or balance between centripetal forces that agglomerate economic activities and centrifugal forces that break up or limit the size of agglomerations.

Ottaviano & Puga (1997) distinguish four mechanisms leading to circular and cumulative spatial concentration of economic activities: labour migration, IO linkages due to intermediate goods, factor accumulation and the combination of intertemporal linkages, history and expectations.

Peripheral regions may lack the critical mass to hold on to an economic activity. Consequently, regional policy may be only temporarily successful in luring economic activity towards the periphery. Even worse, increasing the degree of inter-regional freedom of trade by large infrastructure projects, by product harmonisation, or by any other act that fosters economic integration may have a perverse impact on the periphery since it may become profitable for firms to relocate to the core and serve the peripheral market from there. The NEG combines insights from regional science within a consistent general equilibrium framework. It stands as the only theory within mainstream economics that takes the economics of location seriously.

2.5.5 Capital and Labour Mobility

The relation between growth and agglomeration depends crucially on capital mobility. Without capital mobility between regions, the incentive for capital accumulation and therefore growth might lead to a self-reinforcing unidirectional agglomeration force that moves all economic activity into one location also called a 'catastrophic' spatial agglomeration. In the absence of capital mobility, some results are in fact familiar to the NEG (Fujita et al., 2001): a gradual lowering of transaction costs between two identical regions first has no effect on economic geography but at some critical level induce catastrophic agglomeration. In the model presented in this chapter, in the absence of migration, 'catastrophic' agglomeration means that agents in the south have no more private incentive to accumulate capital and innovate. The circular causality which gives rise to the possibility of a core-periphery structure is common in economic geography models that are characterised by both production and demand shifts which reinforce each other. The production shift takes the form of capital accumulation in one region (and de-accumulation in the other) and the demand shift takes the form of increased permanent income due to investment in one region (and a decrease in permanent income in the other region).

2.5.6 Agglomeration/Dispersion Forces

Three different explanations exist for the existence of spatial agglomerations: one is the natural advantages of different regions, another captures the intuition that agglomeration can occur when firms benefit from production externalities, and the last one describes how market access forces may cause agglomeration. Each type of agglomeration produces different relationships between residential land rents in regions where an industry operates, that industry's productivity, that industry's share of regional employment, and the diversity of regional employment across all industries.

The natural advantages model includes different industries, each of which uses an industry-specific raw material, as well as labour, in its production process. Regions differ in their endowments of these raw materials. These raw materials cannot be transported from one region to another, so firms may find it advantageous to locate in regions with a large endowment of the industry-specific natural resource used in their production process. Thus, natural advantage is manifested in the exogenous pattern of raw material availability; producers located in a region with a large endowment of their required raw material will have an advantage over competitors located in a region with a small endowment of their required raw material, since their raw material input will be available at lower cost.

Next firms benefit from local production externalities, which exist when a firm's production possibilities depend on the actions of other firms located in the same region. Specifically, a firm's production possibilities vary with the employment decisions of other nearby firms. This type of production externality may be the result of knowledge spillovers that occur when employees in different firms have informal contact with each other, or because high turnover in local employment allows a firm to benefit from the knowledge its employees acquired at their previous jobs. Production externalities are typically classified as either localisation externalities or urbanisation externalities. Localisation externalities exist when firms benefit from proximity to similar firms or other firms in the same industry. Urbanisation externalities exist when firms benefit from locating in a diverse location or from proximity to firms engaged in variety of different industries.

A third source of agglomeration is agents' desire for proximity to markets. If goods are costly to transport, then consumers want to be close to as many producers as possible in order to reduce the cost of purchasing goods. At the same time, producers want to be close to as many consumers as possible in order to have the largest possible market for selling their output. Together, these forces generate a circular causality in which agglomeration occurs because consumers and producers want to be close to each other.

2.5.7 Review of the Empirical SCGE Models

2.5.7.1 MIRAGE

MIRAGE is a multi-region, multi-sector CGE model developed by CEPII for trade policy analysis.[3] MIRAGE uses the latest version of the Global Trade Assessment Project (GTAP) global database. The model includes the representation of imperfect competition, product differentiation and FDI. MIRAGE includes a very detailed representation of trade barriers based on the database MAcMaps. The model has been tested using the trade liberalisation between the European Union and its periphery as an example. Imperfect competition in the model follows a Cournot oligopolistic framework. It combines horizontal product differentiation with geographical differentiation. The model is dynamic and uses a recursive setup. The number of operating firms in each of the oligopolistic sectors is determined endogenously in the model. Dynamic part of the model includes the representation of capital reallocation and FDI flows. MIRAGE includes three attractive modelling features that distinguish it from other trade-oriented CGE models: (1) explicit modelling of FDI flows, (2) vertical product differentiation, (3) representation of monetary and non-monetary trade barriers.

[3] See http://www.mirage-model.eu/miragewiki/index.php?title=Accueil Accessed 24.07.13.

2.5.7.2 GEM-E3

GEM-E3 is an applied general equilibrium model for the European Union that has been developed by the consortium led by Technical University of Athens and subsequently used for energy and environmental policy analysis.[4] The model is based on EuroStat data. It includes the representation of markets for goods, services, labour and capital. GEM-E3 represents each EU member state individually and links them with the international trade. It incorporates the representation of various sectors of the economy. The model is recursively dynamic over time and includes the representation of capital accumulation and technological progress. Investment decisions are based on backward-looking expectations. The model has detailed representation of the governmental sector with its major taxes, subsidies and transfers. The focus of GEM-E3 is energy-environmental policies and systems. It includes the representation of emissions and abatement curves.

2.5.7.3 DART

The dynamic applied regional trade (DART) general equilibrium model is a multi-region, multi-sector, recursive-dynamic CGE model that has been developed by the University of Kiel (Springer, 1998). The model uses GTAP database for calibration of its parameters. Each region of the model includes the representation of producers, consumers, government and investments. The world in DART is divided into several regions that are linked by international trade. It is a recursive dynamic model where the time periods are connected by capital accumulation. Dynamics of the model is based on assumptions about economic growth rates and population changes. The output of each sector is produced using the combination of energy with intermediate inputs and primary factors of production including capital, labour and land. Energy use in the model is associated with the greenhouse gas (GHG) emissions.

2.5.7.4 MONASH

The MONASH model has been developed by the Centre of Policy Studies and Impact Project at Monash University, Australia.[5] The model is an extension of the static ORANI model for Australia. MONASH is a multi-regional, multi-sectoral dynamic CGE model system which allows for different choices of the levels of sectoral and regional disaggregation. One can also apply different assumptions with regard to behaviour of economic agents. In particular one can change assumptions about the types of expectations of the economic agents. The model includes the representation of inter-regional transport margin that are required for trade of goods and services in the model. All economic agents make inter-temporal decisions, where firms maximise their discounted profits by making investment decisions and households decide upon the share of disposable income that is spent on consumption and savings. MONASH allows for different types of closures. Depending on

---

[4] See http://ipts.jrc.ec.europa.eu/activities/energy-and-transport/gem-e3 Accessed 24.07.13.
[5] See http://www.monash.edu.au/policy/monmod.htm Accessed 24.07.13.

the type of the model run different assumptions regarding the set of exogenous variables is adopted.

2.5.7.5   SCGE Model for Japan

Toshihiko Miyagi developed the SCGE model for Japan used for the assessment of the indirect economic effects from Tokai–Hokuriku Expressway for the whole of the country (Miyagi, 2001). The aim of the model is to calculate the multiplier effects of the transport investment project at the regional and country levels. Japan is split into nine regions based on the data from inter-regional IO table. The modelling on inter-regional transport flows is based on the use of CES functions.

2.5.7.6   CGEurope Model

CGEurope developed by the University of Kiel is a spatial general equilibrium model for a closed system of 270 regions covering the whole world, with a focus on Europe (NUTS2), and uses the NEG framework (Bröcker et al., 2010). All regions are treated separately and are linked through endogenous trade. The inference method is comparative static, which means that in each model run two equilibriums (benchmark and scenario) are compared. The basis for comparison is a generally understandable indicator, like real income, real GDP and the equivalent variation measure.

In each region resides a set of households, owning a bundle of immobile production factors, which is used by firms for production of goods. Two types of goods are considered: local and tradable. Local goods can only be sold within the region of production, while tradables are sold everywhere in the world, including the own region.

Producers of local goods combine primary factor services, local goods and tradables, using nested Cobb–Douglas technology with region-specific cost share parameters. The output of local goods is assumed to be completely homogeneous, and is produced under constant returns to scale. Firms take prices for inputs as well as for their output as given, and they do not make any excess profits.

Instead of directly selling this output to households or other producers, firms can use it as the only input needed to produce tradables. The respective technology is increasing returns to scale. Tradable goods are modelled as being close but imperfect substitutes, following the Dixit–Stiglitz approach. Different goods stem from producers in different regions. Therefore, relative prices of tradables do play a role. Changes of exogenous variables (transport costs) make these relative prices change and induce substitution effects. For producers of tradables, only input prices are given, while the output price can be set under the framework of monopolistic markup pricing. Due to free market entry, however, profits are driven to zero, as they are in the market for local goods.

Two features that give the CGEurope model its spatial dimension are:

- the distinction of goods, factors, firms and households by location, and
- the explicit incorporation of transaction costs for goods, depending on geography as well as national segmentation of markets.

Two kinds of trade costs are introduced: costs related to geographic distance (transport costs) and costs for overcoming impediments to international trade. The former are modelled under the assumption that transport costs are increasing with distance but at diminishing rate. The change of these costs will constitute the policy scenario.

A new dynamic version of the model has been developed recently under the European Commission's FP6 REFIT project and is presently being tested (Bröcker & Korzhenevych, 2013). The dynamic version includes inter-temporal investment and savings decisions of the households and firms.

Experiences with the CGEurope model show the possibility of a successful implementation of an applied SCGE model for all European regions as well as the possibility of modelling investment decisions in a dynamic way within the model structure. The model in its present formulation has a simple and clear mathematical structure derived from microeconomic theory. The number of sectors represented in the model is limited to only two: tradable and non-tradable, which does not allow for extensive sectoral analysis and limits the type of policies which can be assessed with the model. The focus of the model is on the assessment of transport-related policies.

2.5.7.7   PINGO Model

The Norwegian model PINGO was developed with the aim of providing forecasts for regional and inter-regional goods transports (Ivanova, Vold & Jean-Hansen, 2002). The model uses the assumption of a small open economy with Norway being represented by 19 regions and one rest of the world region, allowing exports and imports. Trade between regions are supported by an explicitly modelled transport sector incurring transport costs. Origin-destination (OD) matrices combined with transport costs from the Norwegian transport model NEMO have been used in order to construct a SAM, used for calibration of the model. The model distinguishes between nine types of production sectors, one service sector and one investment sector where the sectors act according to the assumption of perfect competition. Through interaction with the transport model NEMO, forecasts are made for mode-specific OD matrices, transport costs, transport volumes, etc. PINGO is a static type model, where forecasts are driven by changes in exogenous variables, such as policy variables and projections of regional populations.

2.5.7.8   SCGE Model for the Philippines

Goce-Dakila and Mizokami (2007) have developed an SCGE model for the Philippines. The aim of their model is to identify the most efficient transport infrastructure investment among three alternative transport modes — land, air and water — for five regions in the Philippines. The model utilises the assumption of perfect competition. It includes the representation of seven production sectors with three types of transport services: water, air and land transport. The demand for transport services is derived on the basis on intermediate demand. Each sector produces using two types of production factors: labour and capital. The effects of transport infrastructure investments are simulated as productivity shocks to the three transportation sectors in the model.

2.5.7.9 B-MARIA Model for Brazil

B-MARIA is the SCGE model for Brazil that has been developed and is presently maintained by the University of Sao Paulo (e.g. see Haddad & Hewings, 2005). Then model includes the representation of 27 regions of Brazil and eight different production sectors each producing a specific commodity. The sectors included into the model are: (1) agriculture, (2) manufacturing, (3) utilities, (4) construction, (5) trade, (6) financial institutions, (7) public administration and (8) transport. Besides the production sectors each region includes the representation of the household and the regional government. At the national level the model includes the representation of the federal government and international trade. The model has been calibrated on the data from the inter-regional IO tables for Brazil for the year 1996. The model is static and can operate under two different closer rules: (1) short run and (2) long run. In addition to the assumption of capital immobility the short-run closure of the model also includes fixed regional population and labour supply in combination with fixed regional wage differentials. In the case of the short run, investments of the firms are exogenously fixed. Governmental deficit has been also exogenously fixed in this version of the model.

The latest version of B-MARIA (Haddad & Hewings, 2005) includes increasing returns to scale via a combination of a CES production function with the fixed entry costs to the industry. The model also includes the representation of the Brazilian transport network. The transport network is included in the model via the inclusion of the real transport margins (in contrast to the iceberg costs assumption) into the model. Transport services in the model are produced by regional transport sectors using labour, capital and intermediate goods. Inter-regional freight transport flows generated by the model are further mapped to a geo-coded transport network. The transport network model allows one to calculate the inter-regional transport costs as a function of the freight flows and report them back to the economic part in an iterative manner.

2.5.8 The RAEM Family of SCGE Models

2.5.8.1 RAEM-Light Model

RAEM-Light model is constructed under the assumption of perfect competition and constant returns to scale and hence are not able to capture the NEG features (Koike & Thissen, 2005). The economy consists of multiple regions that are linked by inter-regional trade. Labour and capital are immobile between the regions. Transport margins are modelled based on the iceberg assumption that is a certain share of commodity disappears during transportation from one region to another. RAEM-Light includes the elements of stochasticity. The choice of the origin for the goods purchased by consumers and producers is determined on the basis of logit type of model. This model estimated the probability that consumers and producers will choose to buy goods from a particular region. This probability if influences by regional prices and transportation costs. The total inter-regional trade flow is calculated as the total demand multiplied by the logit probability. The model has been implemented for Japan, the Netherlands and Hungary.

2.5.8.2 RAEM Model

RAEM is the SCGE model for the Netherlands that has been developed by the Netherlands Organisation for Applied Scientific Research (TNO) in cooperation with the University of Groningen. It has also been implemented for Belgium, BeNeLux, Norway, Russia and the European Union (RAEM-Europe). The model fits in the NEG theory. The latest (dynamic) version of the model (RAEM 3.0) for the Netherlands has been developed in a joint project of TNO with Transport and Mobility Leuven during the years 2006–2007 (Ivanova et al., 2007).

The RAEM model includes the representation of the microeconomic behaviour of the following economic agents: production sectors differentiated according to the SBI93 Dutch classification; an investment agent; federal government and an external trade sector. The model is a dynamic, recursive over time, model, involving dynamics of capital accumulation and technological progress, stock and flow relationships and backward-looking expectations. The recursive dynamic structure is composed of a sequence of several temporary equilibriums.

The level of the unemployment benefits, received by the household, depends upon the level of unemployment associated with this particular household type. The voluntary unemployment in the economy is modelled according to the wage curve, which relates the level of the unemployment and the level of the real wages in the economy.

The RAEM model adopts the assumption of the average cost pricing in combination with the assumption of the Dixit–Stiglitz varieties and monopolistic competition between the firms inside each sector. Under the monopolistic competition framework, it is assumed that each sector consists of a number of identical firms, each producing a unique specification of a particular commodity. The model incorporates the representation of the federal government. The governmental sector collects taxes, pays subsidies and makes transfers to households, production sectors and to the rest of the world.

2.5.8.3 RHOMOLO Model of EU DG Regional Policy and JRC-IPTS

RHOMOLO is a regional holistic model for Europe which has been implemented at the level of NUTS2 regions for a number of EU countries (Brandsma, Ivanova & Kancs, 2011). The model integrates economic environmental energy and social dimensions in one unified framework.

RHOMOLO is especially developed for an ex-ante impact assessment of the European Cohesion Policy (ECP). The model can also be used for ex-post impact assessment, other policy simulations and comparison between policy scenarios. RHOMOLO incorporates the following important features:

- Link regions within a NEG framework
- Has inter-temporal dynamic features with main endogenous growth engines including accumulation of knowledge and human capital
- Incorporates public sector interventions
- Incorporates a multi-level governance system

The RHOMOLO model is an equilibrium model with inter-regional trade and location choice based on microeconomics, using utility and production functions with substitution between inputs. It is able to model (dis)economies of scale, external economies of spatial clusters of activity, substitution between capital, labour, energy and material inputs in the case of firms, and between different consumption goods in the case of households. Moreover, monopolistic competition of the Dixit—Stiglitz type allows for heterogeneous products implying variety, and therefore allows for cross hauling of close substitutes of products between regions.

All production activities in the model are associated with emissions and environmental damage. The model incorporates the representation of all major GHG and non-GHG emissions. Emissions in the model are associated either with the use of energy by firms or with the overall level of the firms' outputs.

The general structure of the SCGE model extends to include endogenous growth elements, such as technological progress and human capital accumulation. Development of these two factors is based on the behavioural decisions of households and firms in the model as well as public expenditure. The model contains the representation of two levels of the government, one level representing the central government and the other representing the regional governments, with two levels of the budgetary system.

The model is dynamic, recursive over time involving dynamics of physical and human capital accumulation and technological progress, stock and flow relationships and adaptive expectations. The main model parameters are estimated econometrically using either time series or panel data techniques. The remainder of the model parameters are calibrated on the latest available data for 2007.

2.6   Conclusions and Ideas for Further Research

The present chapter presented a review of present theoretical and empirical approaches to predicting inter-regional freight transport flows. Continuing population and economic growth, trade specialisation and globalisation trends of the last decades have resulted in unprecedented increase in freight transport movements both in terms of volume as well as in terms of average distance. Reduction in costs and improvement in efficiency of freight transportation has acted as the driver of further globalisation, economic growth and contributed to global integration of the capital market and increase in the amount and scope of FDI investments in different parts of the world.

International and inter-regional freight demand is determined by the spatial distribution of production and consumption activities. International freight flows mirror to the large extend the trade flows between the countries.. The main driving forces of freight transport demand include population growth and migration, technological development, specialisation including access to natural resources, agglomeration and dispersion forces. These drivers are quite similar to the drivers of international trade which determines the methodologies that are currently being used for prediction of international and inter-regional demand for freight transportation services.

The main theories that are currently used in order to investigate the spatial distribution of economic activating and inter-regional trade flows include microeconomic theory in particular Von Thuenen and Alonso models, spatial interaction theory in particular gravity model and spatial accounting models in particular multi-regional IO and SCGE models. The present chapter has considered in-depth the three approaches to modelling trade and freight flows: gravity model, multi-regional IO model and finally the SCGE model. Each of the described approaches has its pros and cons that I would like to briefly discuss below.

The main positive feature of the gravity model is that it is estimated econometrically and hence its results can be statistically validated. Unfortunately, there are relatively few applications of the gravity model to detailed commodity trade statistics. Most of the empirical models explain the total flows of trade between the countries without going into the more commodity details. This is partly explained by the quality of the data and the complexity of the econometric methods that are required to deal with missing data and many zero trade flows. The present aggregate level of details used in gravity modelling limits into usefulness for predicting freight transport flows.

Multi-regional IO models with fixed trade coefficients have relatively low data requirements and are easy in implementation. They capture the essence of interregional interactions but ignore the changes in trade patterns as response to changes in transportation costs. The later drawback is largely overcome with the use of random utility-based trade coefficients. They are based on econometric estimations with micro-level data of transport surveys and take into account the impacts of the transportation costs. However, these models ignore the income and supply side effects that could be important in case of major changes in transport infrastructure and/or transport policy.

Both income and supply side effects are taken fully into account in the SCGE models. These models have theoretically sound assumptions and elegant mathematical structure. They are currently becoming more popular among researchers and policy makers as the quality of regional data and the computer power is gradually improving. The major drawback of these models is high implementation costs and impossibility to estimate econometrically all main behavioural equations. The current practice is to estimate econometrically a number of main model parameters which calibrate the rest on the base-year data set.

As you could see from the mathematical derivations presented in the chapter the three strands of modelling have a lot in common. The theoretical structure of the gravity equation is derived on the basis of spatial general equilibrium model and hence fully consistent with theoretical assumptions of SCGE modelling. SCGE models use IO data as the core of its modelling framework. In many cases, they also use Leontief fixed technological coefficients for the calculation of intermediate production inputs.

Given the present speed of data quality improvement and advancements in computational and computer techniques, SCGE modelling can be seen as the future of regional economic modelling and a preferred methodology for predicting inter-regional transport flows and effects of changes in transport policy and transport infrastructure. However, it is important to include well empirically validated and econometrically tested functional forms and parameters into the models in order to make them more robust and reliable.

References

Anderson, S. P., De Palma, A., & Thisse, J. -F. (1987). The CES is a discrete choice model? *Economics Letters, 24*, 139–140.

Banister, D., & Berechman, Y. (2001). Transport investment and the promotion of economic growth. *Journal of Transport Geography, 9*, 209–218.

Bergstrand, J. H., Egger, P., & Larch, M. (2013). Gravity redux: estimation of gravity-equation coefficients, elasticities of substitution, and general equilibrium comparative statics under asymmetric bilateral trade costs. *Journal of International Economics, 89*, 110–121.

Brandsma, A., Ivanova, O., & Kancs, d'A (2011). *RHOMOLO – a dynamic spatial general equilibrium model.* Seville, Spain: JRC IPTS.

Bröcker, J., & Korzhenevych, A. (2013). Forward looking dynamics in spatial CGE modelling. *Economic Modelling, 31*, 389–400.

Bröcker, J., Korzhenevych, A., & Schuermann, C. (2010). Assessing spatial equity and efficiency impacts of transport infrastructure projects. *Transportation Research Part B, 44*, 795–811.

Capros, P., Georgakopoulos, T., Filippoupolitis, A., Kotsomiti, S., & Atsaves, G. (1997). The GEM-E3 model for the European Union: reference manual National Technical University of Athens and others.

Erlander, S., & Stewart, N. F. (1990). *The gravity model in transportation analysis: theory and extensions.* Utrecht, The Netherlands; Tokyo, Japan: VSP.

Fachin, S., & Venanzoni, G. (2012). *IDEM: an integrated demographic and economic model of Italy.* CONSIP S.p.A.

Fujita, M., Krugman, P. R., & Venables, A. J. (2001). The spatial economy: cities, regions, andinternational trade. Massachusetts: MIT Press.

Goce-Dakila, C., & Mizokami, S. (2007). Identifying transport infrastructure investment with maximum impact: a SAM-based SCGE approach. *Journal of the Eastern Asia Society for Transportation Studies, 7*, 376–391.

Haddad, E. A., & Hewings, G. J. D. (2005). Market imperfections in a spatial economy: some experimental results. *The Quarterly Review of Economics and Finance, 45*, 476–496.

Helpman, E., Melitz, M., & Rubinstein, Y. (2008). Estimating trade flows: trading partners and trading volumes. *The Quarterly Journal of Economics, 123*, 441–487.

Hunt, J. D., & Abraham, J. E. (2009). PECAS – for spatial economic modelling: theoretical formulation. Calgary, Canada: HBA Specto Incorporated.

Ivanova, O., Vold, A., & Jean-Hansen, V. (2002). *PINGO: A model for prediction of regional-and interregional freight transport* (version 1). Norway: TOI. (Rep. No. 578/2002).

Ivanova, O., Heyndrickx, C., Spitaels, K., Tavasszy, L. A., Manshanden, W., Snelder, M., et al. *RAEM: version 3.0.* Transport and Mobility Leuven.

Koike, A., Thissen, M. J. P. M. (2005). Dynamic SCGE model with agglomeration economy (RAEM-Light) memo, TNO.

Kratena, K., Streicher, G., Temurshoev, U., Amores, A. F., Arto, I., Mongelli, I., et al. *FIDELIO 1: fully interregional dynamic econometric long-term input-output model for the EU27.* Seville: JRC IPTS.

Miyagi, T. (2001). Economic appraisal for multi-regional impacts by a large scale expressway project. *Tinbergen Institute Discussion Paper*, TI 2001-066/3.

Ottaviano, G. I. P., Puga, D. (1997). Agglomeration in a global economy: a survey of the 'new economic geography'. *CEP Discussion Paper*, 356.

Santos Silva, J. M. C., & Tenreyro, S. (2006). The log of gravity. *The Review of Economics and Statistics, 88*, 641−658.

Sivakumar, A. (2007). Modelling transport: a synthesis of transport modelling methodologies. Imperial College London Working Paper.

Springer, K. (1998). The DART general equilibrium model: a technical description. Kiel Working Paper, 883.

Thissen, M. J. P. M. (1998). A classification of empirical CGE modelling (Rep. No. 99C01). University of Groningen, The Netherlands.

U.S. Department of Commerce, (1997). Regional multipliers: a user handbook for the regional input-output modeling systems (RIMS II).

Suggestions for Further Reading

Arnott, R., Braid, R., Davidson, R., & Pines, D. (1999). A general equilibrium spatial model of housing quality and quantity. *Regional Science and Urban Economics, 29*, 283−316.

Arrow, K. J., Chenery, H. B., Minhas, B. S., & Solow, R. M. (1961). Capital-labor substitution and economic efficiency. *The Review of Economics and Statistics, 43*, 225−250.

Baier, S. L., & Bergstrand, J. H. (2001). The growth of world trade: tariffs, transport costs, and income similarity. *Journal of International Economics, 53*, 1−27.

Bergkvist, E., Westin, L. (1998). Forecasting interregional freight flows by gravity models. Utilising OLS-, NLS- estimations and Poisson-, neural network- specifications. In: *38th Congress of the European Regional Science Association.* Vienna, Austria, August 28−September 1, 1998, session B1.

Bougheas, S., Demetriades, P. O., & Morgenroth, E. L. W. (1999). Infrastructure, transport costs and trade. *Journal of International Economics, 47*, 169−189.

Chen, N., & Novy, D. (2011). Gravity, trade integration, and heterogeneity across industries. *Journal of International Economics, 85*, 206−221.

Chow, J. Y. J., Yang, C. H., & Regan, A. C. (2010). State-of-the art of freight forecast modeling: lessons learned and the road ahead. *Transportation, 37*, 1011−1030.

Comi, A., Delle Site, P., Filippi, F., & Nuzzolo, A. (2012). Urban freight transport demand modelling: a state of the art. *European Transport, 51*, 1−17.

De Jong, G., Gunn, H., & Walker, W. (2004). National and international freight transport models: an overview and ideas for further development. *Transport Reviews, 24*, 103−124.

French, S. (2011). The composition of exports and gravity. FREIT, Working paper.
French, S. (2013). The composition of trade flows and the aggregate effects of trade barriers.
Funk, M., Elder, E., Yao, V., & Vibhakar, A. (2006). Intra-NAFTA trade in mid-south industries: a gravity model. *The Review of Regional Studies, 36*, 205–220.
Gaulier, G., Zignago, S. (2004). The role of proximity and similarity in trade of goods.
Grosche, T., Rothlauf, F., & Heinzl, A. (2007). Gravity models for airline passenger volume estimation. *Journal of Air Transport Management, 13*, 175–183.
Ham, H., Kim, T. J., & Boyce, D. (2005). Implementation and estimation of a combined model of interregional, multimodal commodity shipments and transportation network flows. *Transportation Research Part B, 39*, 65–79.
Hu, D. (2002). Trade, rural–urban migration, and regional income disparity in developing countries: a spatial general equilibrium model inspired by the case of China. *Regional Science and Urban Economics, 32*, 311–338.
Huang, T., & Kockelman, K. M. (2008). The introduction of dynamic features in a random-utility-based multiregional input-output model of trade, production, and location choice. *Journal of the Transportation Research Forum, 47*, 23–42.
Kepaptsoglou, K., Karlaftis, M. G., & Tsamboulas, D. (2010). The gravity model specification for modeling international trade flows and free trade agreement effects: a 10-year review of empirical studies. *The Open Economics Journal, 3*, 1–13.
Khadaroo, J., & Seetanah, B. (2008). The role of transport infrastructure in international tourism development: a gravity model approach. *Tourism Management, 29*, 831–840.
Kockelman, K. M., Jin, L., Zhao, Y., & Ruiz-Juri, N. (2005). Tracking land use, transport, and industrial production using random-utility-based multiregional input.output models: applications for Texas trade. *Journal of Transport Geography, 13*, 275–286.
Miller, R. E., & Blair, P. D. (2009). *Input-output analysis: foundations and extensions*. Cambridge: Cambridge University Press.
Lankhuizen, M. B. M., De Graaff, T., De Groot, H. L. F. (2012). Product heterogeneity, intangible barriers and distance decay: the effect of multiple dimensions of distance on trade across different product categories. *Tinbergen Institute Discussion Paper*, TI 2012-065/3.
Lofgren, H., & Robinson, S. (2002). Spatial-network, general-equilibrium model with a stylized application. *Regional Science and Urban Economics, 32*, 651–671.
Marto Sargento, A. L. (2009). Regional input-output tables and models: interregional trade estimation and input-output modelling based on total use rectangular tables. *Dissertation* Faculdade de Economia, Universidade de Coimbra.
Nagurney, A., & Aronson, J. (1988). A general dynamic spatial price equilibrium model: formulation, solution, and computational results. *Journal of Computational and Applied Mathematics, 22*, 339–357.
Nagurney, A., & Zhang, D. (1996). On the stability of an adjustment process for spatial price equilibrium modeled as a projected dynamical system. *Journal of Economic Dynamics and Control, 20*, 43–62.
Nijkamp, P. (2007). Ceteris paribus, spatial complexity and spatial equilibrium: an interpretative perspective. *Regional Science and Urban Economics, 37*, 509–516.
Nijkamp, P., Reggiani, A., Tsang, W. F. (1999). Comparative modelling of interregional transport flows: applications to multimodal European freight transport. 1999. Vrije Universiteit Amsterdam. Research Memorandum.
Novy, D. (2013). International trade without CES: estimating translog gravity. *Journal of International Economics, 89*, 271–282.
Park, J. Y., Cho, J. K., Gordon, P., Moore, J. E. I., Richardson, H. W., & Yoon, S. S. (2011). Adding a freight network to a national interstate input–output model: a TransNIEMO application for California. *Journal of Transport Geography, 19*, 1410–1422.

Ramirez Grajeda, M. & De Leon Arias, A. (2009). Spatial implications of international trade under the new economic geography approach. MPRA Paper No. 18076. Universidad de Guadalajara.

Russo, F., & Musolino, G. (2012). A unifying modelling framework to simulate the spatial economic transport interaction process at urban and national scales. *Journal of Transport Geography*, *24*, 189–197.

Sargento, A. L. M. (2009). Introducing input-output analysis at the regional level: basic notions and specific issues. *REAL Discussion Papers*, 09-T-4.

Segall, R. S. (1995). Mathematical modelling of spatial price equilibrium for multicommodity consumer flows of large markets using variational inequalities. *Applied Mathematical Modelling*, *19*, 112–122.

Simmonds, D., & Feldman, O. (2011). Alternative approaches to spatial modelling. *Research in Transportation Economics*, *31*, 2–11.

Takahashi, T. (2004). Spatial competition of governments in the investment on public facilities. *Regional Science and Urban Economics*, *34*, 455–488.

Tavasszy, L. A. (2006). Freight modelling – an overview of international experiences. In: *TRB conference on freight demand modelling: tools for public sector decision making*, Washington, DC, September 25–27, 2006.

Tavasszy, L. A., Davydenko, I., Ruijgrok, K. (2009). The extended generalized cost concept and its application in freight transport and general equilibrium modeling. In: *Seminar integration of spatial computable general equilibrium and transport modelling*. The University of Tokyo, SANJO-Hall, Bilateral Joint Seminar under agreement between NWO and JSPS, August 19–20.

Tavasszy, L. A., Ruijgrok, K., Davydenko, I. (2010). Incorporating logistics in freight transportation models: state of the art and research opportunities. In: *12th WCTR*, Lisbon, Portugal, July 11–15.

Tavasszy, L. A., Thissen, M. J. P. M., & Oosterhaven, J. (2011). Challenges in the application of spatial computable general equilibrium models for transport appraisal. *Research in Transportation Economics*, *31*, 12–18.

Truong, T. P., & Hensher, D. A. (2012). Linking discrete choice to continuous demand within the framework of a computable general equilibrium model. *Transportation Research Part B*, *46*, 1177–1201.

Ueda, T., Koike, A., Yamaguchi, K., & Tsuchiya, K. (2005). *Spatial benefit incidence analysis of airport capacity expansion: application of SCGE model to the Haneda project*, Global competition in transportation markets: analysis and policy makingresearch in transportation economics (*Vol. 13*, pp. 165–196). Amsterdam: Elsevier.

Verhoef, E. T. (2005). Second-best congestion pricing schemes in the monocentric city. *Journal of Urban Economics*, *58*, 367–388.

Wang, J. (2012). Estimating regional freight movement in Australia using freight info commodity flows and input-output coefficients. 20th IIOA conference, Bratislava Conference paper.

Wiedmann, T. (2009). A review of recent multi-region input–output models used for consumption-based emission and resource accounting. *Ecological Economics*, *69*, 211–222.

Yang, C. W., Hwang, M. J., & Sohng, S. N. (2002). The Cournot competition in the spatial equilibrium model. *Energy Economics*, *24*, 139–154.

Zhao, Y., & Kockelman, K. M. (2004). The random-utility-based multiregional input–output model: solution existence and uniqueness. *Transportation Research Part B*, *38*, 789–807.

3 Freight Generation and Freight Trip Generation Models

José Holguín-Veras[a], Miguel Jaller[a], Ivan Sánchez-Díaz[a], Shama Campbell[a] and Catherine T. Lawson[b]

[a]Center for Infrastructure, Transportation, and the Environment, and the VREF's Center of Excellence for Sustainable Urban Freight Systems, Department of Civil and Environmental Engineering, Rensselaer Polytechnic Institute, Troy, NY, USA, [b]Department of Geography and Planning University at Albany, Albany, NY, USA

3.1 Introduction

The freight transport system is a key contributor to the vibrancy of local and regional economies. However, in spite of its importance, the functioning of the system is still poorly understood. To a great extent, this is the result of its complexity and challenge of collecting data to characterise it. Some of the elements that make freight such a complex system are the multiplicity of the participating agents, the range of transport modes and geographical areas, and the different measures used to define and measure freight. Some of these factors are at the root of the difference between the freight system and its better understood counterpart, the passenger transport system (Holguín-Veras et al., 2012a). These differences include that freight is entirely passive and require the intervention of agents to conduct loading and unloading activities; there is a wide range of commodities that are being transported; the decisions about mode, route and delivery times are made by different agents and, there is a large difference between the freight demand and the freight traffic that transport it (Friedrich, Haupt, & Noekel, 2003; Holguín-Veras & Thorson, 2000; Ogden, 1992; Ortúzar & Willumsen, 2001).

This chapter provides an overview of a critical aspect of freight demand modelling, which is related to the estimation of the freight generation (FG) (amount of cargo generated) and the freight trip generation (FTG) (number of freight vehicle trips generated) required to transport the freight generated. As discussed later in the chapter, it is important to treat FG and FTG separately because not doing so will lead to erroneous results. This is important because FG/FTG analyses play a key role in the assessment of the traffic impacts produced by land use changes, and in

Modelling Freight Transport. DOI: http://dx.doi.org/10.1016/B978-0-12-410400-6.00003-3

long-term transport modelling exercises. Since the majority of the cargo movements in urban areas are transported by road (e.g. trucks and delivery vans), the remainder of this chapter will discuss FTG based on this mode.

The chapter provides an overview of the literature in Section 3.2, reinterprets FG and FTG from the logistical perspective in Section 3.3, and establishes the key factors to be taken into account when modelling in Section 3.4, while Section 3.5 discusses empirical results from the New York City (NYC) metropolitan area. The chapter ends with the statement of general conclusions.

3.2   Literature Review

FG refers to the production and attraction of cargo, measured by tonnage or volume (e.g. m^3). FTG, in contrast, measures the number of freight vehicle trips that are generated by the transport of FG. Treating FG and FTG as separate concepts is important because while FG is directly correlated with the size of the establishments, FTG may not (Holguín-Veras et al., 2011). This is the result of the role played by the shipment size, which enables large business establishments to receive larger shipments minimally increasing the amount of vehicle trips produced. In this context, a better understanding of the variables driving the generation of freight and freight trips would enable more accurate demand forecasts, and better quantification of the traffic impacts of freight activity. In consistency with the practices in passenger transport modelling, one could subdivide FG and FTG in attractions and productions (Ortúzar & Willumsen, 2001), leading to the concepts of freight attraction (FA), freight production (FP), freight trip attraction (FTA) and freight trip production (FTP).

Different methodologies have been used in FG and FTG modelling. Table 3.1 summarises the advantages and disadvantages of the various techniques (Bastida & Holguín-Veras, 2009; Jong, Gunn, & Walker, 2004). The applications also differ in terms of the dependent and independent variables that have been used, the levels of aggregation and geography, and model structure.

3.2.1   FTG Models

There are a number of publications discussing FTG models, for the most part using constant trip rates and ordinary least squares (OLS) models. Brogan (1980) identified land use policies as the most effective stratification strategy to improve FTG models. Bartlett & Newton (1982) estimated regression models for FTG using total employment, site area, gross floor area and non-office employment as independent variables. Middleton, Mason, & Chira-Chavala (1986) analysed FTG for special land use classes. Tadi & Balbach (1994) estimated trip generation rates for different vehicles types and non-residential land uses using traffic counts. The Quick Response Freight Manual (Cambridge Systematics Inc., 1996) produced FTG rates based on employment to estimate the number of trips at the zonal level, volumes at

Table 3.1 Advantages and Disadvantages of FG/FTG Models

Type of Model	Advantages	Disadvantages
Time series	Require multiple data points, over time, for the same facility. Limited data requirements for independent variables	Little insight into causality and, limited possibility to study policy effects
Trip rates	Simple to calculate	Unable to connect the effect of business size on FTG which may lead to significant errors
	Limited data requirements (zonal data)	Little insight into causality and, limited scope for policy effects
Input−output	Linked to the economy	Need input−output table, preferably multi-regional
	Policy effects could be considered if coefficients are elastic	Need to identify import and export trade flows
		Restrictive assumptions if fixed coefficients
		Need conversion from values to tonnes
Ordinary least squares (regression)	Able to identify relations pertaining to demand generation; can be used not only to forecast future demand, but also to establish the linkage between variables	Violations of the ordinary least square (OLS) assumptions could lead to inaccurate parameters; especially using aggregated data
Spatiat regression	Improves model fit; eliminates problems associated with spatial autocorrelation	Choice of a spatial model depends on actual data and it is hard to pre-determine which structure is more appropriate
Cross classification method	Good regional estimates	May produce errors for traffic impact analyses
	No need for linearity assumption between variables	Independent variables used may not be independent
		Needs a sizeable and detailed database to develop models
MCA	Can overcome the disadvantages of cross classification analyses	May overestimate the future number of FG or FTG if the number of observations by category is not adequately selected

(*Continued*)

Table 3.1 (Continued)

Type of Model	Advantages	Disadvantages
Neural networks	Can produce accurate results; do not need to preselect independent variables; the learning capability of the model can discover complex interactions among independent variables.	Need a sizeable database to develop and calibrate the model

Source: Holguín-Veras et al. (2012a).

external stations and trips between zones. The Federal Highway Administration (FHWA) Guidebook on State Travel Forecasting (Federal Highway Administration, 1999) estimates truck trips based on land use and trip data obtained from travel diaries. Iding, Meester, & Tavasszy (2002) estimated OLS models on data on industrial sites and found that the most appropriate independent variables depend on the industry sector and whether it is an FTA or an FTP model. The ITE Trip Generation Manual (Institute of Transportation Engineers, 2008) contains a compilation of FTG rates for different land uses. Bastida & Holguín-Veras (2009) used OLS models, trip rates and multiple classification analyses (MCA) on disaggregate FTG data to estimate disaggregate FTA rates. They found that commodity type, industry sector and employment are the variables that better predict FTG. Holguín-Veras et al. (2012a) analysed the errors introduced by assuming a constant FTG rate and concluded that it could lead to large estimation errors if it is used to estimate FTG for industry segments where the FTG does not depend on business size. This key issue is further discussed later in the chapter.

Using FTG data for the NYC metropolitan area, Holguín-Veras et al. (2011) estimated disaggregate establishments-level OLS models for FTG, using employment as the independent variable. They found that in 51% of the industry segments analysed, the FTG was constant as it does not depend on business size; for 31% the FTG was a function of a constant and an FTG rate per employee; and the rest of the cases FTG was a constant rate per employee. This provides confirmatory evidence of the need to decouple FG from FTG, and calls into question common practices of using FTG rates per employee. Lawson et al. (2012) estimated and compared disaggregate models for different land use classes using two different land use classification systems. Furthermore, Holguín-Veras et al. (2013) assessed the transferability of these FTG models using external validation data (FTG data from NYC carriers and receivers, New York Capital Region, Mid-West furniture chain and Seattle region grocery stores). They found that the disaggregate models developed by Holguín-Veras et al. (2011) and Lawson et al. (2012) outperformed the models provided by the ITE manual (Institute of Transportation Engineers, 2008) and the Quick Response Freight Manual (U.S. Department of Transportation,

1996; U.S. Department of Transportation, 2007). Sanchez-Diaz, Holguín-Veras, & Wang (2013) studied the role of network characteristics and spatial autocorrelation on FTA. They found that locational variables (e.g. street width and distance to truck routes) influence FTA. Moreover, retail establishments were found to have significant spatial autocorrelation. For an overview of FTG models developed in Europe see Taniguchi & Thompson (2002) and Patier & Routhier (2008). These models are generally based on zonal aggregates or data from surveys conducted at different geographic locations.

Other authors have analysed FTG at special facilities. Studies of port FTG have been conducted by Guha & Walton (1993); Wegmann, Chatterjee, Lipinski, Jennings, & McGinnis (1995); Al-Deek, Johnson, Mohamed, & El-Maghraby (2000); Al-Deek (2001); Holguín-Veras, López-Genao, and Salam (2002); Wagner (2010). For warehouse FTP include the works of DeVries & Dermisi (2008) and Orsini, Gavaud, & Bourhis (2009). In addition to rates and regression models, times series models (Garrido, 2000), input—output (Sorratini, 2000), neural networks (Al-Deek, 2001) and other related models have been used for FTG.

3.2.2 FG Models

Different techniques have been used for FG modelling. Novak, Hogdon, Guo, & Aultman-Hall (2007) used OLS to develop FP models for the United States. They analysed different variable transformation techniques and the implication of spatial regression for FP. Waliszewski, Ahanotu, & Fischer (2004) used commodity-type specific growth rates to estimate FP and FA at the zonal level.

Input—output and spatial computable general equilibrium (SCGE) models are recurrently used to estimate FP and FA at the zonal level (see also Chapter 2 of this volume). Some of the models focusing on regional applications include Sorratini & Smith (2000); Boyce (2002)Hewings, Sonis, & Boyce (2002); Zhao & Kockelman (2004); Al-Battaineh & Kaysi (2005); Giuliano, Gordon, Pan, Park, & Wang (2007). For specific models used in European countries see the SMILE and RAEM (Dutch) models (Oosterhaven, Knaap, Ruijgrok, & Tavasszy, 2001; Tavasszy, van de Vlist, Ruijgrok, & van de Rest, 1998) that estimate FP and FA by linking production and consumption in product chains; the SAMGODS (Swedish) input—output model (Swahn, 2001); the CGEurope (German) SCGE model (Bröcker, 1998); the PINGO (Norwegian) SCGE model (Ivanova, Vold, & Jean-Hansen, 2002); and the integrated regional economic freight model of the United Kingdom (WSP Policy & Research, 2005). Holguín-Veras et al. (2012a) compiled the FG/FTG models reported in the literature in a relational database that is available at Holguín-Veras et al. (2012b).

3.3 Logistical Interpretation of FG/FTG

A key aspect to understand FG and FTG is the production process taking place in the establishment, which is the *raison d'être* of the businesses being there in the first place. The nature of the inputs, the amount of vendors a firm uses to obtain the supplies needed to operate, and the FG and FTG the establishment creates are essentially determined by the type of industrial activity and size of the establishment. One could expect that in general, although large businesses require larger quantities of inputs, they do not necessarily require a more diverse set of inputs, because the production process may be similar to the one at smaller establishments. This conjecture was corroborated with empirical evidence by Holguín-Veras et al. (2012a), where statistical analyses showed that there is no significant relationship between the number of suppliers and business size for a sample of establishments in NYC.

The number of trips generated depends on the logistic decisions of supplier and receiver. In the context of economic specialisation and a competitive market, suppliers have to deliver the supplies in the way specified by the receivers because not doing so may lead to loss of customers, even though the absence of consolidation leads to higher transport costs. As truck trips are indivisible, any delivery or shipment produces one trip whether it transports a full truckload or a small amount of cargo. In this context, supply chain modelling provides an important perspective on FG/FTG. The main reason is, as discussed before, related to the fact that the FG/ FTG is the result of the orders that business establishments place for the supplies needed for their economic activities. As a result, understanding the logic behind these business decisions helps gain insight into FG and FTG.

Generally, the FG is a function of business size as the larger an establishment is, the larger the volume of cargo that arrives and departs from it. However, the situation with FTG is different because FTG is not only impacted by the amount of cargo, but also by the shipment size used during transport. The role of shipment size is quite significant because businesses could increase FG, without increasing FTG, by simply increasing shipment size, changing the type and size of the vehicle or mode used, or all simultaneously. Therefore, an increase in the FG required does not necessarily translate into an increase in FTG. The constant number of inputs required for operations, the indivisibility of truck trips, and the flexibility of increasing shipment sizes challenge the premise that FTG is proportional to business size. The inescapable conclusion is that the flexibility provided by the ability to change shipment size enables large establishments to proportionally generate less FTG than smaller ones.

In terms of logistics decisions, inventory theory, and particularly the economic order quantity (EOQ) model (Harris, 1915; Holguín-Veras et al., 2011), help explain this phenomenon. The EOQ computes the optimal combination of shipment size and delivery frequency that minimises the total logistics cost TC (transport plus inventory costs) associated with transporting a given amount of cargo (see Chapter 5 of this volume for additional discussions of the EOQ model and choice of shipment size). In the simplest case, TC is equal to the summation of: (Eq. 3.1) the cost of placing the order (fixed cost A per order, for order set-up); (Eq. 3.2) the transport costs, which include a fixed transport charge K per trip (often specific to mode, covering loading/unloading, drayage, among others) and a variable cost of transport per unit c; and (Eq. 3.3) the inventory cost per unit h. TC can be expressed as:

$$TC = (A + K)\frac{D}{Q} + h\frac{Q}{2} + cD \tag{3.1}$$

where Q is the shipment size, thus D/Q is the total number of shipments to be transported. From this equation, the EOQ model establishes that the optimal shipment size Q^*, and the optimal time between orders, T^*, can be derived as:

$$Q^* = \sqrt{\frac{2(A + K)D}{h}} = \sqrt{\frac{2(\text{Set-up cost} + \text{transportation cost})(\text{demand per unit time})}{\text{inventory cost}}} \tag{3.2}$$

$$T^* = \sqrt{\frac{2(A + K)}{hD}} = \sqrt{\frac{2(\text{Set-up cost} + \text{transportation cost})}{(\text{inventory cost})(\text{demand per unit time})}} \tag{3.3}$$

From T^*, one could find the optimal frequency f^* as:

$$f^* = \frac{1}{T^*} = \sqrt{\frac{hD}{2(A + K)}} = \sqrt{\frac{(\text{Inventory cost})(\text{demand per unit time})}{2(\text{set-up cost} + \text{transportation cost})}} \tag{3.4}$$

These equations highlight the differences between FG and FTG. For instance, while the FTG equals the number of vehicles used to transport Q^* times the delivery frequency f^*; FG is the demand per unit time (D). The models show that increases in FG have a proportionally smaller increase in FTG, as the increase in FG is managed by smaller increases in shipment size and delivery frequency. Therefore, a four-time increase in FG will result only in the doubling of the FTG. It is important to mention that the EOQ model provides a good estimate of the shipment sizes observed in real life (see Combes (2012) and Chapter 5 of this volume). Moreover, it is expected that modes or vehicle types with high set-up and transport costs will be used to transport larger shipments less frequently. In addition, for the cases of non-bundled shipments, it would be optimal to select the vehicle type with capacity closer to the shipment size. In essence, to transport the amount of cargo needed by large establishments, businesses increase both shipment size and delivery frequency. As a result, the increase in FTG is typically less than proportional to the underlying increase in FG with respect to a smaller establishment. This is illustrated in Figure 3.1.

Empirically, Figure 3.2 shows the number of deliveries per employee for a sample of (53) establishments in the wholesale trade industry in the New York Metropolitan Area. The figure shows that the number of deliveries per employee for small businesses is about six times the number of deliveries for large businesses.

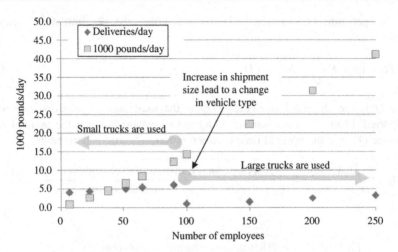

Figure 3.1 Conceptual relationship between FG, FTG and business size.
Source: Holguín-Veras et al. (2012a).

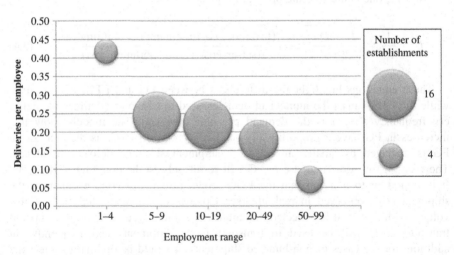

Figure 3.2 FTA of establishments in the wholesale trade industry.
Source: Holguín-Veras et al. (2012a).

3.4   Factors to Take into Account when Estimating or Applying FG and FTG

The research conducted by José Holguín-Veras et al. (2012a, 2012b) clearly suggests that the quality of the estimates of FG/FTG depends on how the data are categorised and aggregated, and the models estimated. This section discusses these important factors.

3.4.1 Classification Systems

Classification systems are useful to group establishments that are likely to have similar FG/FTG patterns. Among other benefits, classification systems enable simple models like the ones typically used for FTG analyses, to increase the statistical quality of the estimates produced, and provide a convenient way to organise the models so that they match local land use ordinance. The classification systems used for FG and FTG could be classified into economic-based and land-use-based. The former classifies establishments on the basis of the industry sector they belong to, while the latter classifies them depending on land use characteristics.

Economic-based classification systems used in the United States include the Standard Industrial Classification (SIC), and the North American Industry Classification System (NAICS). The first attempts at formulating standardised economic classification systems dates back to the 1930s resulting from the need to collect and analyse industry statistics, to avoid ambiguities and to be able to make comparisons across regions and agencies. The first complete edition of the SIC was produced in 1939, with a final revision almost 40 years later in 1987 which included 1004 different industries. The SIC was then replaced in the 1990s due to, among other issues, concerns about its inefficient coverage of emerging service sectors. This led to the development of the NAICS in 1997 which is the current industrial classification system in use in the United States, covering a larger number of sectors than the SIC, though numerous legacy data still use SIC.

Land use classification systems have been used for many years to support land use zoning and planning. However, the variety of activities and characteristics makes it difficult to create a land use classification system appropriate for all communities. The first attempt to formulate a nationwide land use classification system was the Standard Land Use Coding Manual (SLUCM) in the 1970s. This system was updated in the 1990s, giving rise to the Land-Based Classification System (LBCS). A unique aspect of the LBCS is that it seeks to classify land use in four different dimensions: activity, function, structural character, site developer character, and ownership (American Planning Association, 1994). Although the LBCS offers great flexibility, it failed to take roots and only a handful of cities adopted it. Instead, most cities and municipalities adopted their own zoning ordinances to regulate the size and use of land and buildings, such as the City of New York Zoning Resolution. For a more detailed description of the different classification systems see Holguín-Veras et al. (2012a).

Recent research has shed light into the performance of these classification systems as the foundation for FG/FTG modelling. The results show that economic classification systems are significantly better than land-use-based classification systems for FG/FTG modelling (Holguín-Veras et al., 2012a). Specifically, the best models were found when combining an economic measure of business size, such as employment, and an economic classification system, such as SIC or NAICS. There are good reasons for that to be the case. To start with, employment is an input to the economic process taking place in the establishment. Thus, under competitive conditions the larger the employment, the larger the output. If the inputs and outputs are physical, one could expect that FG would be related to employment, and to a lesser extent, FTG. In contrast, the amount of space occupied by an establishment is not necessarily an input to the production process (only in a handful of sectors, such as agriculture and mining, land could be considered an input); more appropriately, the space is a constraint. Moreover, most land use classification classes are very general in nature (e.g. 'commercial') and, as a result, tend to cluster a disparate collection of establishments with very different economic characteristics. The internal heterogeneity of such classes significantly reduces the ability of the models to produce statistically solid results.

3.4.2 Level of Aggregation

Models can be classified in two main groups depending on the type of data used. Aggregate models are able to provide average estimates over a group using an average of many individual data points, while disaggregate models use less but more detailed data points to explain individual behaviour. Advantages of disaggregate models include having the potential to provide more temporally and spatially consistent models; increased efficiency in data usage as they require less data points as input; ability to utilise the inherent variability in the data, which is lost in aggregate models; and a lower chance of having biases in the model due to correlation that exist between aggregate units (Ortúzar & Willumsen, 2001). Disaggregate models could be used to produce aggregate results but attention has to be paid in using correct aggregation procedures. This important subject is discussed next.

3.4.3 Aggregation Procedures

In order to conduct appropriate spatial aggregation at the zonal level using disaggregate models, it is important to take into consideration the mathematical structure of the disaggregate models. Although there is a wide range of cases, the discussion here uses as an example employment-based models. However, the aggregation procedures will hold for other models of similar structure (Holguín-Veras et al., 2011). In general, the total FTG of a zone, F, would be the aggregate of the FTG of the n independent establishments, f_i, in the zone.

$$F = \sum_{i=1}^{n} f_i \tag{3.5}$$

Due to the linear nature of f_i, there are only three possibilities: a constant α per establishment (Type S); a rate β per employee (Type E); or a combination of a constant and a term that depends on employment, E (Type C). Table 3.2 shows the appropriate aggregation procedure for each type of model.

In essence, for the Type S models, the total FTG is the product of the unit FTG (α) and the number of establishments (n). For the Type E models, the total is estimated by the product of the FTG rate (β) and total employment (E^*). Total

Table 3.2 Spatial Aggregation Procedures for Disaggregated Models

Type	Model Type	Model Structure	Aggregation Procedure
S	FTG constant per establishment	$f_i = \alpha$	$F^S = \sum_{i=1}^{n} \alpha = n\alpha$
E	FTG rate per employee	$f_i = \beta E_i$	$F^E = \sum_{i=1}^{n} \beta E_i = \beta \sum_{i=1}^{n} E_i = \beta E^*$
C	FTG is a constant and a term that depends on employment level	$f_i = \alpha + \beta E_i$	$F^C = \sum_{i=1}^{n} (\alpha + \beta E_i) = n\alpha + \beta \sum_{i=1}^{n} E_i = n\alpha + \beta E^*$

employment is estimated by the summation of the individual employment, E_i, of the n establishments in the zone. Finally, the aggregate FTG for Type C models is the product of the total number of establishments and the constant (α) added to the product of the total employment (E^*) and the FTG rate (β).

Furthermore, for zones that exhibit the presence of establishments with different FTG patterns, that is, their FTG is determined by different model structures, the aggregate FTG for the different groups of establishments must be estimated following the appropriate procedure from Table 3.2 and then add the individual totals. Mathematically:

$$F = F^S + F^E + F^C \tag{3.6}$$

3.5 Case Study: FTG in the New York City Metropolitan Area

This section provides a summary of the research conducted by the authors, for the most part centered in the NYC metropolitan area, so that the readers could get an idea about the FTG patterns in a large metropolis. The data used in the research were collected via two surveys, one for carriers and another for receivers. Survey questions were concentrated in the areas of company attributes, operational and FTG. The sample selected was comprised of receiver companies in Manhattan and carrier companies located in selected counties of New York and New Jersey from different industry sectors. The approaches used in the estimation process were ordinary least squares (OLS) and MCA. The OLS models considered the functional form in which FTG is a function of the total number of employees per establishment. The dependent variables were freight trips production (FTP) and freight trips

Table 3.3 FTA (Deliveries/Day) Models by NAICS

Economic Classification System: NAICS	Obs.	Constant	Rate per Employee	RMSE
23 − Construction*	25	2.160		1.364
31, 32 and 33 − Manufacturing*	51	2.831		2.791
31 − Food, beverage, tobacco, textile, apparel, leather and allied product manufacturing	21	2.400		1.295
32 − Wood, paper, printing, petroleum and coal products, chemical, plastics, non-metallic and mineral product manufacturing	10	4.420		5.483
33 − Metal, machinery, computer, electronic, electrical, transportation, furniture and misc. manufacturing	20	2.490		2.483
42 − Wholesale trade*	117	2.272	0.069	3.655
44 and 45 − Retail trade*	98	3.070	0.063	4.054
44 − Motor vehicle, furniture, electronics, building material, food and beverage, health, gasoline and clothing stores	69	2.458	0.132	4.298
45 − Sporting goods, hobby, book and music stores	29	2.724		4.352
72 − Accommodation and food*	56	1.307	0.081	3.091

*Group models.
Source: Adapted from Holguín-Veras et al. (2012a).

attraction (FTA). The use of such linear models gives rise to one of the three types of models (S, E and C) described in the previous section depending on the level of significance of the parameters. The measure used to evaluate the efficiency of the model in estimating FTP and attraction is the root mean square error (RMSE). Table 3.3 shows the disaggregate models for FTA considering business size for different industry segments. The results indicate that 60% of the models were a constant trip rate per establishment while the remaining models being a constant and rate per employee. These suggest, as discussed in Holguín-Veras et al. (2011), that using constant trip rates per employee may not be the most appropriate approach to estimate FTA.

To gain further insight into the relation between FTG and business size, the authors selected those industries that exhibited a combination of a constant per establishment and a term based on employment (Type *C* models) to estimate MCA models. Table 3.4 shows the resulting MCA models for these industries. These estimates indicate that, on average, those establishments in the retail trade industry (NAICS 44) receive one more delivery than those in the accommodation and food service (NAICS 72) and wholesale trade (NAICS 42) industries.

Table 3.4 MCA Rates for FTA (Deliveries/Day)

		Economic Classification System: NAICS		
		42: Wholesale Trade	44: Retail Trade	72: Accommodation and Food
Employees*	1–10	2.443	3.543	1.902
	11–20	3.341	4.442	2.801
	21–30	5.685	6.785	5.144
RMSE		3.658	4.197	3.355

*For establishments with more than 30 employees results were not statistically significant.
Source: Adapted from Holguín-Veras et al. (2012a).

Similarly for FTP, Table 3.5 shows the resulting disaggregate models for the different NAICS codes. For FTP, 30% of the models are Type S, another 30% are Type E, and the remaining 40% are Type C. These results show, yet again, that a large number of industry sectors have a constant FTA that does not depend on business size.

Table 3.6 shows the MCA results for the different industries. These results indicate that establishments in the construction (NAICS 23) industry averaged approximately one more trip than those in the manufacturing (NAICS 32) industry. Establishments in the transport and warehousing (NAICS 42 and 48) industries both averaged at least one more trip than establishments in the retail trade (NAICS 44) industry.

For a comparison between estimating methodologies and for a description of models estimated for other industry classification systems and land use categories see Holguín-Veras et al. (2011), Lawson et al. (2012) and Holguín-Veras et al. (2013).

3.5.1 Transferability of FG and FTG Models

An important consideration for FG and FTG modelling is how transferable these models are. This is because, quite frequently, FG/FTG models are used for low budget studies, such as traffic impact analysis mandated by local codes to ensure that proposed developments do not produce undue negative effects on the transport networks. Transferable models are particularly useful when there is a lack of data to produce separate models.

However, in spite of its importance, the amount of research conducted to assess transferability of transport models is very small. In order to help fill this void, Holguín-Veras et al. (2013) assessed how well the models estimated were able to estimate FTG at locations different than NYC. These analyses were conducted using econometric techniques, and applying the models estimated to compute the corresponding estimation errors. The results obtained are discussed next.

Table 3.5 FTP (Trips/Day) Models by NAICS

Economic Classification System: NAICS	Obs.	Constant	Rate per Employee	RMSE
23 − Construction*	9		0.068	1.586
31, 32 and 33 − Manufacturing*	28	2.214		3.599
31 − Food, beverage, tobacco, textile, apparel, leather and allied product manufacturing	13	2.846		4.990
32 − Wood, paper, printing, petroleum and coal products, chemical, plastics, non-metallic and mineral manufacturing	7		0.023	0.648
33 − Metal, machinery, computer, electronic, electrical, transportation, furniture and misc. manufacturing	8	1.750		1.639
42 − Wholesale trade*	124	1.755	0.036	5.094
44 and 45 − Retail trade*	9		0.161	6.485
44 − Motor vehicle, furniture, electronics, building material, food and beverage, health, gasoline and clothing stores	5	0.993	0.021	0.237
48 and 49 − Transportation and warehousing*	157	2.718	0.038	4.811
48 − Air, rail, water, truck, transit, pipeline, scenic and sightseeing, and support activities	153	2.725	0.038	4.005

*Group models.
Source: Adapted from Holguín-Veras et al. (2012a).

Table 3.6 MCA Models for FTP (Trips/Day)

		Economic Classification System: NAICS				
		23 − Construction	32 − Wood, Paper, Petroleum, Coal, Chemical	42 − Wholesale Trade	44 − Motor Vehicle, Furniture, Electronics, Food and Beverage	48 and 49 − Transportation
	1−20	2.424	1.303	2.946	1.685	3.381
	21−40	1.727	0.606	2.564	1.303	2.998
Employees	41−60	2.061	0.939	3.283	2.023	3.718
	61−80	4.061	2.939	2.764	1.504	3.199
	>80	5.121	4.000	7.609	6.348	8.043
RMSE		1.074	0.934	4.650	0.618	5.219

Source: Adapted from Holguín-Veras et al. (2012a).

To conduct the econometric analysis of transferability, Holguín-Veras et al. (2013) joined FTG data from 30 different cities and included binary variables to represent the different areas. The resulting pooled data was used to estimate models that included the locational binary variables. The results indicated that these binary variables, in most cases, were not statistically significant. The implication is that there are no location specific effects that, as a result, the models were transferable (at least within the US context).

The second technique used to assess transferability entailed a comparative analysis of the models available in the literature (Beagan, Fischer, & Kuppam, 2007; Institute of Transportation Engineers, 2008) and the FTG models reported in Campbell, Jaller, Sanchez-Diaz, Holguín-Veras, & Lawson (2011), National Cooperative Freight Research Program (2012), and Holguín-Veras et al. (2011). The results produced by Holguín-Veras et al. (2013) are shown in Table 3.7.

Table 3.7 shows that, in all but one case, the NCFRP 25 models perform much better than the ITE and the QRFM models. Moreover, the RMSE and the gap in performance between NCFRP 25 and the QRFM model tend to increases with employment, as predicted in Holguín-Veras et al. (2011). However, the geographical origin of the validation data does not seem to affect the RMSE, thus suggesting some degree of transferability.

3.5.2 Spatial Effects on Freight Trips Attraction

An additional enhancement to FTG modelling is the incorporation of locational variables and spatial autocorrelation to account for spatial effects. Sanchez-Diaz et al. (2013) show that the inclusion of locational variables (e.g. width of front street, distance to truck routes and distance to large traffic generators) can improve the explanatory power of FTA models by as much as 80%. Another spatial effect affecting FTA is spatial autocorrelation; these effects can be isolated or controlled using spatial econometrics. As shown in Sanchez-Diaz et al. (2013), retail establishments are particularly affected by spatial autocorrelation. According to the empirical results, in the retail industry, FTA is negatively affected by proximity between establishments. In other words, proximity between retail establishments is associated to lower FTA.

3.6 Conclusion

The chapter provided an overview of the concepts of FG and FTG and the corresponding modelling processes. The chapter focused on road transport mode for FTG since trucks and other delivery vehicles are the dominant mode for urban goods movements. The discussions establish the importance of treating FG and FTG as two different concepts. While FG is the amount of cargo generated by an establishment, FTG is the number of freight vehicle trips required to transport the cargo. In competitive markets, larger businesses are expected to produce proportionally more FG than

Table 3.7 Estimation Errors for NCFRP 25, QFRM and ITE Models

| Classification | Description | Validation Data | | | NCFRP 25 | | Model Applied | | | |
| | | Sample Size | Mean Employment | Data Set | Model | RMSE | QFRM | | ITE | |
							Model	RMSE	Model	RMSE
NAICS 72	Accommodation/ food	5	5.8	NYS-CR	1.307 + 0.081*E	1.26	1.206*E	6.51	n/a	n/a
ITE 816	Hardware/paint stores	8	10.0	NYC	0.369*E	1.67	1.206*E	1.99	53.21*E Trucks: 2%	2.04
LBCS	Activity restaurants	5	5.8	NYS-CR	2.488	1.93	1.206*E	6.51	n/a	n/a
ITE 890	Furniture stores	12	10.0	NYC	3.769	2.22	1.206*E	4.58	12.19*E Trucks: 5%	3.39
LBCS	Function retail	13	8.9	NYS-CR	3.682	2.55	1.206*E	22.46	n/a	n/a
ITE 890	Furniture stores	58	8.9	MW-FC	3.769	3.42	1.206*E	5.60	12.19*E Trucks: 5%	1.25
ITE 860	Wholesale markets	102	17.2	NYC	2.272 + 0.069*E	3.66	1.206*E	12.23	8.21*E Trucks: 30%	11.66
SIC 56	Apparel/ accessory	10	10.2	NYS-CR	1.889 + 0.187*E	4.05	1.206*E	23.25	n/a	n/a
NAICS 44	Grocery stores	7	15.3	SR-GS	2.458 + 0.132*E	4.10	1.206*E	32.06	n/a	n/a
SIC 58	Eating/drinking places	5	5.8	NYS-CR	4.307 + 0.081*E	4.14	1.206*E	6.51	n/a	n/a
SIC 52	Building materials stores	6	18.8	NYS-CR	5.260	4.42	1.206*E	36.14	n/a	n/a
LBCS	Activity goods	21	13.0	NYS-CR	2.588 + 0.067*E	4.56	1.206*E	23.81	n/a	n/a

SIC 54	Food stores	8	19.5	NYS-CR	$3.000 + 0.288*E$	5.09	$1.206*E$	26.04	n/a
NAICS 44	Grocery stores	30	78.0	NYC-GS	$2.458 + 0.132*E$	7.08	$1.206*E$	41.73	n/a
NAICS 44	Retail trade	21	55.0	NYS-CR	$3.451 + 0.153*E$	8.02	$1.206*E$	23.42	n/a
LBCS	Function grocery	8	19.5	NYS-CR	$0.217*E$	13.89	$1.206*E$	26.04	n/a

Notes: SIC, standard industrial classification; NAICS, North American classification system; LBCS, land-based classification standard; NYS-CR, New York state capital region data; NYC, NYC data; NYC-GS, NYC grocery stores data; MW-FC, mid-west furniture chain data; SR-GS, Seattle region grocery stores data; grocery stores, furniture stores and LBCS models only consider attraction; ITE models estimate total passenger trips and truck trips proportion; E, employment; and RMSE, root mean squared error.
Source: Adapted from Holguín-Veras et al. (2013).

small ones, thus FG increases with business size. This is not necessarily the case for FTG, since logistics decisions on shipment size, vehicle (and mode) choice and frequency of distribution come into play when determining the number of truck trips generated. Furthermore, the indivisibility of truck trips leads to a situation where the transport of a small shipment takes the same number of truck trips than a large one. Thus, small businesses tend to produce proportionally more FTG than large ones.

As a result, it is important to try to capture these logistics patterns through the use of disaggregate (establishment level) models. Furthermore, the research conducted by the authors Holguín-Veras et al. (2012a) suggests that the quality of estimates depend on how the data are categorised and aggregated, and the model estimated. In this context, the chapter describes economic-based and land-use-based classification systems used for modelling purposes, and discusses the advantages of economic-based systems based on results from previous research. The chapter also provides an overview of the different methodologies used in the literature to estimate FG and FTG and discusses the advantages and limitations of each. Since disaggregate models are preferred, the chapter stresses the importance of using the appropriate aggregation procedures when producing aggregate estimates and describes the aggregation procedures for different functional forms of FG/FTG models estimated using ordinary least square (OLS).

Using the NYC Metropolitan area as a case study, the authors show a number of FTG models estimated using ordinary least squares (OLS) and MCA. The models were estimated for different industry segments using employment as the independent variable. The results show that using business size (employment) alone may not be the most appropriate approach to estimate FTG, since 60% of the attraction and 40% of the production models resulted in constant rates per establishment. These results were validated with external data. In addition, the chapter discusses the need to evaluate the transferability of estimated FG/FTG models following previous research conducted by the authors Holguín-Veras et al. (2013). In addition, the chapter discusses the analyses conducted to assess the role of locational variables and spatial effects. The results indicate that these effects could greatly improve the explanatory power of FTG models.

The general and more detailed discussions provided in this chapter provide a complete overview of the FG and FTG estimation process. It is expected that these contribute to the improvement of the state of the art and practice of freight demand modelling.

References

Al-Battaineh, O., & Kaysi, I. (2005). Commodity-based truck origin-destination matrix estimation using input-output data and genetic algorithms. *Transportation Research Record: Journal of the Transportation Research Board, 1923*(1), 37–45.

Al-Deek, H. M. (2001). Comparison between neural networks and multiple regression approaches for developing freight planning models with specific applications to seaports. In: *80th annual meeting of the transportation research board*, Washington, DC.

Al-Deek, H. M., Johnson, G., Mohamed, A., & El-Maghraby, A. (2000). *Truck trip generation models for seaports with container-trailer operations. 79th annual meeting of the Transportation Research Board*. Washington, DC: National Academy Press.

American Planning Association. (1994). Toward a standardized land-use coding standard. Retrieved 15.03.10, from <www.planning.org/lbcs/background/scopingpaper.htm>.

Bartlett, R. S., & Newton, W. H. (1982). Goods vehicle trip generation and attraction by industrial and commercial premises. Transport and Road Research Laboratory. Berkshire, England. ISSN: 0305−1293.

Bastida, C., & Holguín-Veras, J. (2009). Freight generation models. *Transportation Research Record: Journal of the Transportation Research Board, 2097*(1), 51−61.

Beagan, D., Fischer M., & Kuppam, A. (2007). Quick response freight manual II. FHWA-HOP-08-010 EDL No. 14396.

Boyce, D. (2002). Combined model of interregional commodity flows on a transportation network. In G. J. D Hewings, M. Sonis, & D. Boyce (Eds.), *Trade, networks and hierarchies*. Berlin, Germany: Springer.

Bröcker, J. (1998). Operational spatial computable general equilibrium models. *Annals of Regional Science, 32*, 367−387.

Brogan, J. D. (1980). Improving truck trip-generation techniques through trip-end stratification. *Transportation Research Record, 771*, 1−6.

Cambridge Systematics Inc. (1996). Quick response freight manual. Final report of the Federal Highway Administration.

Campbell, S., Jaller, M., Sanchez-Diaz I., Holguín-Veras J., & Lawson, C. (2011). Comparison between industrial classification systems in freight trip generation (FTG) modeling. In: *91st annual meeting of the transportation research board*. Washington, DC.

Combes, F. (2012). Empirical evaluation of economic order quantity model for choice of shipment size in freight transport. *Transportation Research Record, 2269*, 92−98.

DeVries, J. B., & Dermisi, S. V. (2008). Regional warehouse trip production analysis: Chicago metro analysis. Illinois Center for Transportation. Chicago, Illinios. Report No. FHWA-ICT-08-025. ISSN: 0197−9191.

Federal Highway Administration (1999). *Guidebook on statewide travel forecasting*. Center for Urban Transportation Studies.

Friedrich, M., Haupt T., & Noekel, K. (2003). Freight modeling: data issues, survey methods, demand and network models. In: *10th annual international conference on travel behavior research*, Lucerne.

Garrido, R. A. (2000). Spatial interaction between trucks flows through the Mexico−Texas border. *Transportation Research A, 33*(1), 23−33.

Giuliano, G., Gordon, P., Pan, Q., Park, J., & Wang, L. (2007). Estimating freight flows for metropolitan area highway networks using secondary data sources. *Networks and Spatial Economics, 10*(1), 73−91. Available from http://dx.doi.org/10.1007/s11067-007-9024-9.

Guha, T., & Walton, C. M. (1993). Intermodal container ports: application of automatic vehicle classification system for collecting trip generation data. *Transportation Research Record, 1383*, 17−23.

Harris, F. (1915). *Operations and costs*. Chicago, IL: A.W. Shaw.

Hewings, G., Sonis, M., & Boyce, D. (2002). *Trade, networks, and hierarchies: modeling regional and interregional economies*. Berlin, Germany: Springer.

Holguín-Veras, J., Jaller, M., Destro, L., Ban, X., Lawson, C., & Levinson, H. (2011). Freight generation, freight trip generation, and the perils of using constant trip rates.

Transportation Research Record, *2224*, 68–81. Available from http://dx.doi.org/ 10.3141/2224-09.

Holguín-Veras, J., Jaller, M., Sanchez-Diaz, I., Wojtowicz, J., Campbell, S., Levinson, D. M., et al. (2012a). Freight trip generation and land use. *National Cooperative Freight Research Program*, *19*. from <http://onlinepubs.trb.org/onlinepubs/ncfrp/ ncfrp_rpt_019.pdf>.

Holguín-Veras, J., Jaller, M., Sanchez-Diaz, I., Wojtowicz, J. M., Campbell, S., Lawson, C. T., et al. (2012b). NCFRP 25 freight generation and freight generation models Database. Retrieved 13.05.13, from <http://transp.rpi.edu/~NCFRP25/FTG-Database. rar>.

Holguín-Veras, J., López-Genao, Y., & Salam, A. (2002). Truck-trip generation at container terminals. *Transportation Research Record*, *1790*, 89–96.

Holguín-Veras, J., Sánchez-Díaz, I., Lawson, C., Jaller, M., Campbell, S., Levinson, H. S., et al. (2013). Transferability of freight trip generation models. *Transport Research Record*.

Holguín-Veras, J., & Thorson, E. (2000). Trip length distributions in commodity-based and trip-based freight demand modeling: investigation of relationships. *Transportation Research Record*, *1707*(1), 37–48.

Iding, M. H. E., Meester, W. J., & Tavasszy, L. A. (2002). Freight trip generation by firms. In: *42nd European congress of the European regional science association*. Dortmund, Germany.

Institute of Transportation Engineers (2008). *Trip generation: an ITE informational report*. Washington, DC: Institute of Transportation Engineers.

Ivanova, O., Vold, A., & Jean-Hansen, V. (2002). PINGO a model for prediction of regional and interregional freight transport. Institute of Transport Economics. Oslo, Norway. Report No. 578/2002. ISSN: 0802–0175.

Jong, G., Gunn, H., & Walker, W. (2004). National and international freight transport models: overview and ideas for future development. *Transport Reviews*, *23*, 1.

Lawson, C., Holguín-Veras, J., Sánchez-Díaz, I., Jaller, M., Campbell, S., & Powers, E. (2012). Estimated generation of freight trips based on land use. *Transportation Research Record*, *2269*, 65–72. Available from http://dx.doi.org/10.3141/2269-08.

Middleton, D. R., Mason, J. M., Jr, & Chira-Chavala, T. (1986). Trip generation for special-use truck traffic. *Transportation Research Record*, *1090*, 8–13.

National Cooperative Freight Research Program (2012). NCFRP 25: freight trip generation and land use. *NCHRP report* from <http://onlinepubs.trb.org/onlinepubs/nchrp/ nchrp_rpt_739.pdf>.

Novak, D. C., Hogdon, C., Guo, F., & Aultman-Hall, L. (2007). Nationwide freight generation models: a spatial regression approach. *Networks and Spatial Economics*, *10*. Available from http://dx.doi.org/1007/s11067-008-9079-2.

Ogden, K. W. (1992). *Urban goods movement: a guide to policy and planning*. Brookfield, VT: Ashgate Publishing Company.

Oosterhaven, J., Knaap, T., Ruijgrok, C. J., & Tavasszy, L. A. (2001). On the development of RAEM: the Dutch spatial general equilibrium model and its first application. In: *41st European regional science association conference*, Zagreb.

Orsini, D., Gavaud, O., & Bourhis, P. L. (2009). Logistics facilities impacts on the territory. In: *2009 European transport conference*, London: AET.

Ortúzar, J. D., & Willumsen, L. G. (2001). *Modelling transport*. New York, NY: John Wiley and Sons.

Patier, D., & Routhier, J. (2008). How to improve the capture of urban goods movement data? In: *8th International conference on survey methods in transport*, Annecy, France.

Sanchez-Diaz, I., Holguín-Veras, J., & Wang, C. (2013). Assessing the role of land-use, network characteristics, and spatial effects on freight trip attraction. In: *Transportation research board 92nd annual meeting*, Washington, DC.

Sorratini, J., & Smith, R. (2000). Development of a statewide truck trip forecasting model based on commodity flows and input-output coefficients. *Transportation Research Record: Journal of the Transportation Research Board, 1707*(1), 49−55.

Sorratini, J. A. (2000). Estimation statewide truck trips using commodity flows and input-output coefficients. *Journal of Transportation and Statistics, 3*(1), 53−67.

Swahn, H. (2001). The Swedish national model system for goods transport.

Tadi, R. R., & Balbach, P. (1994). Truck trip generation characteristics of nonresidential land uses. *ITE Journal, 64*(7), 43−47.

Taniguchi, E., & Thompson, R. G. (2002). Modeling city logistics. *Transportation Research Record, 1790*, 45−51.

Tavasszy, L. A., van de Vlist, M., Ruijgrok C., & van de Rest J. (1998). Scenario-wise analysis of transport and logistic systems with a SMILE. In: *WCTR conference*, Antwerp.

U.S. Department of Transportation (1996). Quick response freight manual.

U.S. Department of Transportation. (2007). Quick response freight manual II. <http://www.ops.fhwa.dot.gov/freight/publications/qrfm2/qrfm.pdf>.

Wagner, T. (2010). Regional traffic impacts of logistics-related land use. *Transport Policy*. Available from http://dx.doi.org/10.1016/j.tranpol.2010.01.012.

Waliszewski, J., Ahanotu, D., & Fischer, M. (2004). Comparison of commodity flow forecasting techniques in Montana. *Transportation Research Record: Journal of the Transportation Research Board, 1870*(1), 1−9.

Wegmann, F. J., Chatterjee, A., Lipinski, M. E., Jennings, B. E., & McGinnis, R. E. (1995). *Characteristics of urban freight systems*. FHWA.

WSP Policy & Research (2005). The EUNET 2.0 Freight and Logistics Model—Final report, regional pilot for economic/logistic methods. UK Department for Transport. London, England. DfT contract ppad 9/134/24.

Zhao, Y., & Kockelman, K. M. (2004). The random-utility-based multiregional input-output model: solution existence and uniqueness. *Transportation Research Part B: Methodological, 38*(9), 789−807.

4 Distribution Structures

Hanno Friedrich[a], Lóránt Tavasszy[b] and Igor Davydenko[b]

[a]Technical University Darmstadt, Germany
[b]TNO, Delft and Delft University of Technology, The Netherlands

4.1 Introduction

The transport of goods does not always follow the direct route from production to consumption. Often, goods take detours and are routed via warehouses or transhipment centres on their way. This is done to bundle goods for transport or to store goods in warehouse locations. Cost savings due to economies of scale in transport and storage are an important driver, as are the service benefits of having inventory close to the customer. In this chapter, we focus on the use of warehouses in the distribution chain. Looking at the general tendency of increasing proportion of high value goods and smaller lot sizes this phenomenon is especially important.

The explicit consideration of distribution structures in freight transport models seems recommendable out of several reasons: first the 'detours' represent a significant part of freight transport. Second, distribution structures are important to explain spatial patterns of freight flows; a region might not have production or consumption but attract freight transport because of its logistics function, accommodating warehouses. Third, the warehouse structures and the routing of flows represent a logistics decision and thus represent possibilities of actors to react to changes in policy or infrastructures that are the object of investigation for the freight transport modeller. And finally, these structures comprise the opportunities for synergies through bundling. Thus, they are important to derive the costs of freight transport.

In this chapter, we present alternative methods to include these structures in freight transport models. We distinguish between models at the disaggregate or micro level and the aggregate or macro level. First, we take a look at the micro level to understand how these structures emerge in detail. We identify their drivers, taking the perspective of an individual firm's logistics network and introduce the typical normative modelling approaches for sub-problems. Also we provide an example of using these approaches (optimisation heuristics) for describing the emergence of structures in disaggregate freight transport models. Then, we move to the macro level, describing the problems of modelling these structures in aggregate freight transport models. Thus we try to build the bridge between the micro and macro view. Three approaches for modelling the structures on the macro level will be introduced: gravity modelling for O/D flows,

Modelling Freight Transport. DOI: http://dx.doi.org/10.1016/B978-0-12-410400-6.00004-5

discrete choice model for distribution structures and hypernetwork models. We conclude the chapter with a brief summary and an outlook on future research topics.

4.2 The Micro Level

At the detailed (micro) level, transport demand consists of commodity flows between locations (establishments) that can be described by the amount (measured in weight, volume and value) of a commodity, transported from one location to another, over a certain time period.

Those locations can be:

- Production locations that are the sources of the commodities.
- Points of sale where commodities are sold to other actors. Often, from this point onwards commodities are transported in very small quantities, for example, in private cars.
- Consumption locations that are the final destination of the commodities. These can again be production locations where the commodity is used as an input for producing another commodity, but also points of final consumption like restaurants or private homes.
- Warehouse locations where commodities are stored, picked and packaged (picking describes the collection of goods from shelves in warehouses).
- Transhipment locations where commodities are reloaded to other vehicles, bundled, or unbundled.

Warehouse and transhipment locations are the result of logistics decisions. For logistics decisions, individual commodity flows are the input for which the optimal distribution structures, supply paths, lot sizes, tours and routes are determined. The decision on distribution structures includes (Pawelleck, 1996):

- Number of echelons, meaning how many layers the hierarchy of warehouses has, common levels are central (CDC) and regional (RDC) distribution centres.
- Number of warehouses/transhipment points on each echelon.
- Allocation of source or destination locations to warehouses/transhipment points.
- Location of warehouses/transhipment points.

Implicitly the decisions on distribution structures include decisions on supply paths, meaning the locations a commodity passes through from source location (production) to destination (consumption) location, lot sizes of commodities transported, tours of the transporting vehicles and routes of the vehicles on the transport infrastructure.

There are different actors that decide on distribution structures, each having a certain scope for decision, meaning the set of flows that is taken into account for the logistics decisions. The most important actors are:

- Production companies, looking at the flows from their production locations to the location of their customers.
- Wholesale companies or retail chains that consider flows of commodities they are trading, originating from locations of many different producing companies and destined to consumption locations or points of sale.

Table 4.1 Logistics Costs Categories with Main Components

Categories	Transport Costs	Storage Costs	Ordering and Handling Costs	Risk Costs
Main cost components	• Driver • Fuel • Vehicle • Infrastructure use • Capital cost during transit	• Capital • Building and equipment • Land • Energy	• Handling • Picking and packaging • Ordering • Order processing	• Out of stock • Loss and damage • Spoilage • Obsolescence

• Logistics companies that also consider flows from locations of many different producing companies to consumption locations or points of sale, with the difference that they do not own the commodities.

The decisions are taken based on flows considered and the objectives of these logistics actors. In logistics dominant objectives are costs and service level (e.g. transport time or product availability). In recent times, many discussions indicate that besides these economic objectives also the other sustainability dimensions 'social' and 'environmental' are becoming more important. The most important drivers and developments behind commodity flows, costs and service levels, will be discussed in the following, after giving a short overview on relevant cost categories as a basis for discussion.

Common logistics cost categories are given in Table 4.1. Detailed discussions on logistics cost categories and their application in transport modelling can be found in many publications, examples are Beuthe, Vandaele, & Witlox (2004), Park (1995), Blauwens, De Baere, & Van de Voorde (2012)) or Friedrich (2010).

Transport costs are all costs caused by the transport of goods between locations. Inbound costs per tonne-km (tkm) or pallet-km (pkm) to a warehouse are usually lower than outbound costs from a warehouse. This is mainly due to lower transport capacity utilisation on distribution tours compared to main runs, smaller vehicles and regional or urban instead of interregional roads causing more time and fuel consumption.

Storage costs include all costs necessary to keep goods at a location. A significant part of storage costs are fixed, at least for a certain time period, these are especially costs for land, building and equipment. Costs for ordering and handling include all costs of activities within the warehouse or transhipment point. Especially in countries with high labour costs all costs for manual handling, picking or packaging of goods play a very important role. Picking describes the collection of goods from shelves in warehouses. Risk costs are all costs of unintended events that might happen. Often they are connected to the service level, for example, having a high availability of goods may cause spoilage while a low availability of goods may cause out of stock situations.

These are the main cost categories currently considered by logistics actors. In the future, sustainability aspects may become more important, and thus also external costs, for example, damages caused by emissions, may play a role. This is already the case for public investments.

4.2.1 Drivers and Their Developments

This section gives an overview of drivers for distribution structures by explaining their possible impact on logistics cost categories and the form of the distribution structures. The relationships described are only rough tendencies that must be verified in detail in the respective situation. Figure 4.1 shows the five categories of drivers described. Besides demand and supply (of goods transported) that are the basis for decisions, there are characteristics of goods handled and the logistic system in place as well as external resources used which might influence the distribution structures.

4.2.1.1 Demand for Goods

The first category covers all drivers connected to demand. Main drivers here are the volume of demand in the delivery points, their distribution in space, the volatility of demand over time and the required lead time. Lower volumes or a lower density of delivery points in space lead to higher (outbound) transport costs. This causes a tendency to more regional warehouses in order to be closer to the delivery points and avoid outbound transport costs. In this context, the current development of increasing e-commerce is especially interesting. A distribution system for delivery to private households would have warehouses or transhipment facilities close to the consumers as it is case for courier, express and parcel (CEP). This is different to many traditional distribution structures in retailing. Volatility in demand and the requirement of short lead times leads to more stocks in warehouses to have a high availability. This leads to higher storage costs and thus to a tendency of fewer warehouses. If the volatility is independent between the delivery points (and the same good is delivered) fluctuations can be balanced by a more central structure, leading to a tendency for additional (central) echelons. Smaller lead time on the other hand requires regional stocks to have a short transport time. This tendency is the opposite of the tendency of fewer warehouses due to higher storage costs.

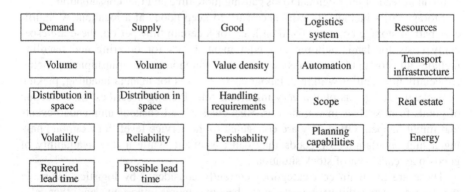

Figure 4.1 Drivers for distribution structures.

4.2.1.2 Supply of Goods

As drivers of demand, drivers of supply are volume of supply per delivery point and distribution in space. However, since inbound costs are much lower the influence is smaller. Instead of volatility and required lead time, one would rather speak of reliability and possible lead time of supply. Both, lower reliability and longer lead time, lead to more stock in the warehouses and thus to an increase in storage costs and a tendency to fewer warehouses. Again, if the same good is supplied and fluctuations through reliability are independent they can be balanced.

4.2.1.3 Goods' Characteristics

Characteristics of goods have a high influence on logistics processes. Besides the value density of goods, special handling requirements and perishability can have an important influence. High value density causes high capital costs of stocks and thus a tendency to smaller transport lot sizes. For the warehouse structure this means higher storage (higher capital cost) as well as higher transport costs. Special handling requirements lead to higher handling costs and the need of special handling equipment. It thus becomes harder to bundle goods with different requirements in the same logistics system. The general tendency is that this leads to fewer echelons and warehouses within a logistics system, but more warehouses overall.

4.2.1.4 Logistic System

For logistics systems automation of logistics activities, the change in planning scope and planning capabilities are relevant drivers. Automation especially can be found within warehouses; this reduces variable storage costs but increases the necessary investments. The tendency for warehouse structures can be different. An example of changes in planning scope is growing logistics systems like the replacement of national by European distribution systems. This may lead to new centres of gravity of demand and supply within the logistics system in space and thus to new locations. Also, a better usage of the logistic system could be the consequence which would result in fewer warehouses. The planning capabilities today change through more powerful IT systems and the availability of data. This can increase the efficiency of all processes and may influence all costs, transport costs can be lower because of better usage or handling costs can be reduced through improved processes in the warehouse, the overall tendency for the distribution system is difficult to predict.

4.2.1.5 Resources

Finally, resources used by logistics systems can be a driver of distribution structures. This includes the availability and costs of transport infrastructure, real estate for logistics locations and energy. Higher costs or a lower service level of infrastructure increases transport cost and therefore cause a tendency to more warehouses to avoid outbound transport. Higher real estate costs lead to higher storage fixed costs and thus a tendency to fewer warehouses. And higher energy prices also lead mainly to higher transport costs and the tendency to more warehouses.

4.2.2 Micro-Level Normative Models

As mentioned before the determination of a distribution structure can include the number of warehouses, warehouse levels, warehouse locations and allocation of consumption points, production locations or customers to warehouses. Each of these problems is highly dependent on the others. Especially, if several of these sub-problems are handled together, finding the optimal solution is very complex and heuristics are used to solve the problems approximately. In the following the most common approaches for the problems are outlined for each individual problem as well as for some combinations.

4.2.2.1 Number of Warehouses

Isolating the number of warehouses problem is difficult, since it highly depends on the other problems like possible locations of warehouses and allocations of drains to warehouses. If, for example, an additional warehouse is added optimal locations of all warehouses change. Tempelmeier (1980, p. 50–57) gives a very detailed problem definition including all cost components. But for the sake of simplicity we want to describe shortly a model of Geoffrion (1979) which catches the most important drivers and also matches quite well with the discussion of drivers in the last section.

Geoffrion (1979) proposes a formulation that does not need data on locations. He assumes an equally distributed demand in a plane, described by a demand density d and the considered area A. The distances to suppliers and consumption points are estimated, based on average geometric distances. He arrives at a relatively simple formula for the optimised number of warehouses:

$$n^* = 0.332 \times F \times \left(d \times \frac{c_2}{w_{\text{fix}}} \right)^{2/3} \times \left(1 - \left(\frac{c_1}{c_2} \right)^2 \right)^{1/2} \tag{4.1}$$

where:

n^* = optimal number of warehouses
F = served area (km^2)
d = demand density in area (Pallets/km^2)
c_1 = inbound transport cost rate (EUR/(km and pallet))
c_2 = outbound transport cost rate (EUR/(km and pallet))
w_{fix} = fixed warehouse costs (EUR (Euro))

Even though this model is simplified in many ways, it makes some dependencies clear, that have been discussed earlier in this chapter. The number of warehouses increases with the service area, with the demand density and with the outbound cost rate; it decreases with the warehouse fixed costs. The variable service costs do not appear in this final formula, since it is assumed that they do not differ by warehouse size and thus do not have an influence. Also, obviously, no differences in stock levels are considered. Thus the model could for example not explain the difference in the number of warehouses between a discounter and a full service retailer in food retailing. The influence of the number of warehouses on stock levels was analysed in detail by Toporowski (1996, p. 82 ff.). Assuming the basic lot size model, he calculates cycle stock (the amount of goods consumed between two deliveries) for n compared to one warehouse. He concludes that the cycle stock increases by a factor of n when changing from one to n warehouses. His analysis shows how significant the influence of stock level is on the decision of the number of warehouses. In terms of costs, an increased stock level is reflected in additional costs for stock positions and capital costs.

4.2.2.2 Warehouse Levels

Now, we focus on the isolated warehouse level problem. Often, this problem can be limited to the question if a second warehouse level (central warehouses) should be established. Therefore, we refer to an approach of Bartholdi and Hackmann (2008, p. 73 ff). Their approach originates from the problem of what articles should be placed in an additional forward area in a warehouse. In this forward area, picking is cheaper and therefore it is recommendable to put those articles in the forward area that promise the highest cost savings. Given is a volume V of the forward area, the volume v_a that is needed to put article a ($a = 1 \ldots A$) into the forward area and the potential saving s_a of putting article a into the forward area. Thus, maximising the total savings S, the optimisation problem can be written as:

$$\text{Max } S = \sum_{a=1}^{A} s_a \tag{4.2}$$

$$\sum_{a=1}^{A} v_a \leq V$$

$$v_a \geq 0, \ \forall \, a \in 1 \ldots A$$

If this idea is transferred to the question of a second (central) warehouse level, it has to be considered which articles should be put into the central warehouse. Instead of setting a certain volume as given, it can be analysed for which article a centralisation would result in savings in variable costs. These savings have to be compared to the additional costs of a central warehouse (C_{add}). Thus, it is beneficial to introduce a central warehouse level, if:

$$\sum_{a=1}^{A} \max\{s_a; 0\} - C_{\text{add}} > 0 \tag{4.3}$$

The savings in variable costs result from lower stock levels and less inbound transport costs.

4.2.2.3 Warehouse Locations

The basic 'warehouse location' (or 'facility location') problem in literature includes the location, allocation and number problems. It will be discussed in the last paragraph of this section. At this point the focus is on the location problem only. If the location can be chosen freely in the plane (no discrete alternatives), the problem is known as 'Steiner−Weber' problem (Domschke & Drexl, 1996, p.167 ff): Given are the locations of the consumption points i ($i = 1 \ldots I$) in form of coordinates (h_i; v_i) and each consumption point has a demand of X_i. Then the (transport) optimal location (h^*; v^*) for the warehouse is where the total distance between the warehouse and the locations is minimised:

$$\text{Min} \sum_{i=1}^{I} X_i \sqrt{(h-h_i)^2 + (v-v_i)^2} \tag{4.4}$$

This function cannot be solved analytically, but Miehle (1958) defines an iterative procedure that follows the Steiner−Weber model. The 'mental model' behind this approach (also called weights and strings model) describes a physical apparatus in form of the table which represents a map. A hole is made at each consumption point with a string going through, holding a weight at the end that corresponds to the demand. All strings are fixed at a ring on the table. The position where the ring comes to a hold is the optimal position. As a starting solution (first iteration in the Miehle procedure), the centre of gravity can be taken:

$$(h_{\text{gr}}; v_{\text{gr}}) = \left(\frac{\sum_{i=1}^{I} h_i X_i}{\sum_{i=1}^{I} X_i} ; \frac{\sum_{i=1}^{I} v_i X_i}{\sum_{i=1}^{I} X_i} \right) \tag{4.5}$$

This location often is already close to the above described minimum. For the following iterations the new coordinates (h; v) are determined by differentiating (4) with respect to a and b, setting the result equal to zero and solving by a and b.

In many cases, however, discrete alternatives are given; also distances often do not correspond to the Euclidian distance but depends on the road network. Therefore, the solutions generated can only be taken as approximation, and locations close to this optimal point have to be checked.

4.2.2.4 Allocation of Consumption Points

Finally, the allocation problem of consumption points to warehouses corresponds to the classic transport problem (Neumann & Morlock, 1993, p. 325 ff.): Given are $i = 1 \ldots n$ warehouses and $j = 1 \ldots m$ consumption points, each consumption point needs the quantity of b_j and each warehouse has the capacity to supply the quantity of

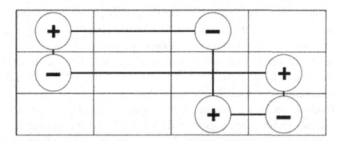

Figure 4.2 Schematic view of MODI approach in a transport tableau.
Source: Simplified from Neumann & Morlock (1993, p. 333).

a_i (total demand equals total supply capacity). The cost matrix c_{ij} represents the transport costs per unit and the matrix x_{ij} the quantity transported between i and j. The objective of the transport problem is to find x_{ij} that minimise total transport costs:

$$\text{Min TC} = \sum_{i=1}^{n} \sum_{j=1}^{m} c_{ij} x_{ij}$$

$$\text{s.t.} \sum_{i=1}^{n} x_{ij} = b_j \ (j = 1 \ldots m)$$

$$\sum_{j=1}^{m} x_{ij} = a_i \ (i = 1 \ldots n)$$

$$x_{ij} \geq 0 \ (i = 1 \ldots n, j = 1 \ldots m)$$

This is a classical linear optimisation problem that can be solved with the simplex method. A special form of the network simplex method is the MODI (Modified Distribution) method. In a transport tableau, representing the flows from the warehouses to the consumption points, cycles are determined that improve costs and change (increase and decrease) flows by a certain quantity so that constraints stay fulfilled (Figure 4.2). This method is a very efficient way to solve the transport problem (Neumann & Morlock, 1993, p. 337). Details on the derivation of this method can be found in Domschke (1981).

This description of the allocation problem corresponds to the case of capacitated warehouses, meaning that warehouses have a limited capacity. The incapacitated case is trivial, since customers can just be allocated to the nearest warehouse.

4.2.2.5 Combined Problems and Solution Procedures

The facility (or plant) location problem and the warehouse location problem in literature are two very closely related problem types that include three of the above described problems: number problem, location problem and allocation problem. The warehouse location problem (Baumol & Wolfe, 1958) describes the problem of placing warehouses between the production and customer locations, the facility location problem describes the problem of placing facilities (plants or depots) to serve customer locations (see for example Sharma & Berry (2007) for a differentiation between warehouse and facility location problem).

Facility location problems can be differentiated by various characteristics; one example is whether facilities are capacitated (Aikens, 1985) for an overview and classification). The simple capacitated facility location problem can be formulated as:

$$\text{Min} \sum_{i=1}^{n} \sum_{j=1}^{n} c_{ij}x_{ij} + \sum_{j=1}^{n} f_i z_i$$

$$\text{s.t.} \sum_{i=1}^{n} x_{ij} = b_j \ (j = 1 \dots m)$$

$$\sum_{j=1}^{m} x_{ij} \leq a_i z_i \ (i = 1 \dots n)$$

$$x_{ij} \geq 0 \ (i = 1 \dots n, \ j = 1 \dots m)$$

$$z_i \in \{0, 1\} \ (i = 1 \dots n)$$

where:

c_{ij} = cost matrix for transport from facility i to consumption point j
x_{ij} = quantity delivered from facility i to consumption point j
f_i = cost of facility i
z_i = 1 if plant is established, 0 otherwise
b_j = demand of consumption point j
a_i = capacity of facility i

At the first glance, the description looks very similar to the transport problem. It differs in the additional consideration of fixed cost f_i for warehouses and binary variables z_i expressing that warehouse locations only represent potential warehouses. Also, the first constraint is formulated more generally to include the case that less than the capacity of the warehouse is used as well.

Nevertheless, this problem is far more complex, it includes the location and the number problem. Finding optimal solutions for large-scale problems does need a lot processing capacity (Beasley, 1988). Heuristics are therefore used to generate good (not optimal) solutions (see Green, Kim, & Lee (1981) for examples).

If the problem of the number of warehouses is separated out the resulting problem is reduced to location and allocation. Despite the simplification, the problem stays complex. Domschke & Drexl (1996, p. 183) propose for the practical application a heuristic from Cooper (1972) that solves the allocation and location problem in turn and thus continuously improves the solution. A similar heuristic was proposed in the area of cluster analysis by Späth (1975), called KMEANS-principle. Both Cooper and Späth show through tests that this approach produces very good results.

4.2.2.6 Heuristic for the Overall Warehouse Structure Problem

In this section, we describe an optimisation heuristic that solves the overall warehouse structure problem including all four component problems. It has been implemented by Friedrich (2010) and is combination of the before mentioned methods for the individual problems. Figure 4.3 gives an overview on the heuristic. In a first phase, the number of warehouses and the warehouse levels are optimised. Then locations of warehouses and allocations of stores to warehouses are determined in a second phase.

The optimisation in the first phase is done by full enumeration, meaning that costs for all possible warehouse structures are calculated. Different kinds of central levels are distinguished by the weekly frequency (fr_{CW}) of transports between central and regional warehouse level. The costs considered within the calculation

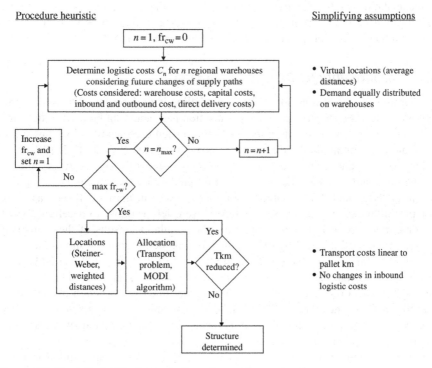

Figure 4.3 Heuristic for the determination of a warehouse structure.

include warehouse costs, capital costs, inbound and outbound costs as well as costs for direct delivery. All these costs are dependent on the structure. The calculation considers all changes caused by a 'new' warehouse structure. Besides new lot sizes for inbound transports this includes possible changes of supply paths. It has to be noted that the heuristic uses some simplifying assumptions to keep the problem solvable. For the determination of the number of warehouses these are the assumptions of virtual locations and equal demand for all warehouses. The locations are assumed to be distributed equally in space, thus, average distances can be used for calculation. Equal demand of warehouses is assumed to avoid a combinatorial problem which needs more time to be solved.

The detailed modelling of this first part differentiates the heuristic from standard problems in operations research like the warehouse location problem or the facility location problem. Contrary to these problems, warehouse costs are modelled in a very detailed way.

The allocation and location problem are solved integrated in an iterative procedure. The choice of locations corresponds to the Steiner—Weber problem. The allocation problem corresponds to the standard transport problem in operations research and can be solved by the MODI algorithm. By repeatedly solving the two problems, the solution is constantly improved. The heuristic terminates if no improvement can be reached through changing warehouse locations or allocation of stores. The implicit assumption, taken in this part, is that inbound costs do not change and thus, no recalculation of the first part is necessary. Also the procedure assumes that the minimisation of tonne kilometres corresponds to the minimisation of costs. More details on both phases of the heuristic can be found in Friedrich (2010).

4.2.3 Applicability for Descriptive Purposes

The purpose of freight transport demand modelling is to describe reality and not to optimise. Thus, a core question is if optimisation heuristics can be used to describe real decisions. The heuristic described before has been used within a logistics simulation model mapping the food retailing sector in Germany (Friedrich, 2010).

For the simulation model data of 31 logistics systems of food retailing companies was available. Five of them were used for calibration the remaining 26 for validation of the model. For calibration some parameters in the model, such as truck cost per kilometre or warehouse fixed cost, were left variable to a certain extent and determined by calibration. Comparing the simulation results of the remaining 26 systems showed the validity of the simulation model. Figure 4.4 shows the overall results of the model. There is a difference of 13 (205 simulated and 213 existing) in the total number of warehouses and an average distance of 60 km between the simulated and the real warehouse. Considering that the resolution of the model was NUTS (Nomenclature of Territorial Units of Statistics) 3 this seems a very good result, meaning that the algorithms (using the calibrated parameters) seem to explain the behaviour of the actors.

For the remaining differences there could be two reasons. First, simplifications have been made in the overall simulation model as well as in the heuristic

Simulated locations   Existing locations

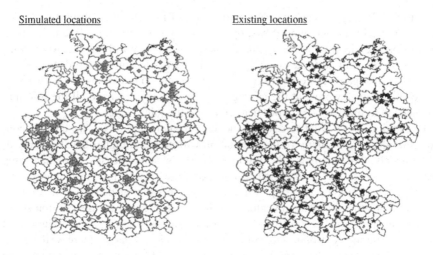

Figure 4.4 Simulated and existing locations of warehouses in German food retailing (Friedrich, 2010).

described; two examples are the assumption of the same size of warehouses within a structure and the green field approach meaning that historical locations of warehouses were not considered. In reality the sizes of warehouses within a structure may deviate and the structures develop over time. The second reason may be that the actors may have different objective functions or may not be as rational as assumed by the heuristic, resulting for example from a lack of information. This is a classical challenge in transport modelling and could be addressed by adding elements like error terms. This could be the next step for model development in this area.

4.3   From Micro to Macro Level

4.3.1   Challenges in Aggregate Models

If we wish to describe and analyse all inventory structures in a whole city, region, state or continent, the micro approach described above will become difficult to apply. Mainly, this is because of the poor availability of data at this detailed level across the entire population of firms and shipments. Most companies do not keep data in sufficient scope and detail to allow a comprehensive micro modelling exercise to be made. As in the previous example, much of the data will have to be acquired by hand. In the future, the data availability situation may change to the better, as more supply chain processes become supported by ICT (Information and Communications Technology) systems. Eventually, however, even if such data is made available by firms, it will be proprietary, difficult to obtain completely and impossible to publish. This leaves us with a formidable challenge. Aggregate statistics of freight transport typically do not include characteristics of logistics operations. Usually we cannot resort to statistics of inventories, transport costs distinguished by in- or outbound flows, or flows by type of origin or destination (e.g. production site, distribution centre and terminal). And even if these data were available, many relevant attributes of freight shipments would not be recorded or would need to be disregarded in order to be able to represent commodities at a higher spatial level. In other words, at the aggregate level we may not have all the information needed to make models work that we use at the micro level.

In addition, if we want to develop models for aggregate flows, we have to consider that the heterogeneity of the population will be much higher than for individual products. For example, the product attributes value density and packaging density determine the number and location of warehouses, through inventory and handling costs. Goods with high value and high packaging density will result in high costs at warehouses for inventory and handling, pushing the distribution structure to become more centralised. If variations for these attributes are significant within a commodity group, and volumes are sufficiently large, there will not be one dominant spatial structure, but several ones, for different product groups. The implication is that we will need to account for diversity in inventory structures, and explain this diversity using the right product attributes. Aggregate choice models are able to deal with heterogeneity in decisions and variations in underlying variables. If this heterogeneity cannot be modelled explicitly, however, commodity groups will need to be as homogeneous as possible. Approaches to formulate homogeneous commodity groups for freight models according to relevant commodity attributes (named 'logistics families'; see Burmeister, 2000 and Tavasszy, 2000) aim to address this problem.

Similar issues are found in the representation of the supply side, the actual inventory networks. As aggregate data superpose networks at the firm level (that each have their own unique rationale) to networks at regional level, the patterns of individual firms are lost and replaced by more aggregate patterns. These aggregate patterns might be different and exhibit a different relation with e.g. transport costs or distance than the underlying firm level patterns. This implies that we cannot be sure anymore that we are modelling the firm's real considerations (be it in a normative or descriptive sense). Instead, we are describing an aggregate logic which might not exist at the firm level. Such aggregation problems are universal in aggregate modelling (i.e. not unique for inventory structures) and widely discussed in the literature (Ortúzar & Willumsen, 2011). Aggregating networks will also lead to loss of information from the perspective of interdependence between networks at the firm level. Especially, where firms outsource their logistics, decisions on inventory structures will be made for several firms at the same time, or partially aggregate flows (or 'meso scale' flows; see Liedtke (2009). Moving from a firm to a regional level will mean that the consideration of this interdependence is lost and replaced by more simplified assumptions of the nature of regional level flows. At the aggregate level we will have to assume that decisions are made in a similar way for entire regions or region pairs – possibly disaggregated, however, by company type or size, or even product type or shipment size – but not by individual firms, let alone interdependent firms.

In the following, we sketch alternative ways to deal with the problem of inventory structures in aggregate models. The modelling problem is framed in the usual aggregate modelling language — that of the four-step transport model. We provide three different ways that are theoretically possible and report on experiences from the literature with each of the three.

4.3.2 Aggregate Modelling of Inventory Structures

4.3.2.1 The Aggregate Modelling Problem

The representation of inventory structures in an aggregate, descriptive model of freight transport is framed as follows.

Let T_{ij} be the flow of trade T (in tonnes/year) between regions i and j. Typically, the origins i will represent production locations, and destinations j the locations where these goods are consumed. The table of trade flows between all these regions we call the P/C matrix. As explained in Chapter 2 of this volume, P/C matrices result from applying multi-regional spatial economic models, such as MRIO (Multi Regional Input Output) or SCGE (Spatial Computable General Equilibrium) models. Because of the existence of intermediate warehouses, these models do not produce the appropriate spatial picture of transport flows, however. In the modelling described below, we aim to include exactly this step.

By inserting warehouses as intermediate origins and destinations at regions k, we convert this P/C matrix into a matrix of pure transport movements, also known as the origin—destination matrix or O/D matrix. In a simple system, of at the most one distribution centre between P and C, the P/C flows are now split into three different underlying O/D matrices: one to represent flows T'_{ij} that go directly between i and j and two to represent those that go indirectly, via a warehouses in regions k: T_{ik} (the P/W flows) and T_{kj} (the W/C flows). Added together they form the O/D table that, when assigned to the transport network, will describe the traffic flows.

The question we treat in this section is how to describe the combined trade and transport system using aggregate models. Conventional models assume that these differences between O/D and P/C do not exist, i.e. assume only one table, which may either be a trade or a transport table. Depending on the approach chosen, they either (1) assign P/C flows directly to the transport networks, or (2) derive an O/D matrix from traffic flows and try to relate this to economic aggregates o producing and consuming regions. The first approach will lead to an underestimation of the amount of traffic, as the additional flows created by indirect movements are neglected. The second will lead to problematic elasticities as the assumption will be made that warehouse-related trips will respond in the same way to economic changes as production or consumption-related trips.

There are several alternative approaches to take into account both tables explicitly. Depending on the approach chosen, more or less detail in terms of the logistics considerations can be put into the model. In the following sections, we discuss:

- Gravity modelling for O/D flows
- Discrete choice model for inventory structures

• Hypernetwork models

Besides providing a brief rationale and mathematical formulation, we review examples of such models from the international literature insofar available and discuss the empirical results obtained.

4.3.2.2   Gravity Modelling Approach

Even if the roots of the well-known gravity model lie in the analogy with Newtonian physics, several theoretical interpretations are available that lend it to modelling trade of commodities (see Chapter 2 of this book). The interpretation that comes closest to our problem is that of fixed price differences and transport costs between regions, inducing trade where the net margin of trade is positive.

If we want to apply this interpretation to the inventory network problem, we have to assume that warehouse regions are also trading points, with product prices allowing to represent their attractiveness as a distribution location, that is as destination for different production regions, and as origin for different consumption regions. It is by this analogy with trade that the following model can develop:

$$T_{ik} = p_i^* q_k^* r_{ik}$$
$$T_{kj} = p_k^* q_j^* r_{kj}$$
$$T_{ij} = p_i^* q_j^* r_{ij}$$

where T, i, j and k are as before and p, q and r are parameters for respectively origin and destination regions in the three O/D tables for transport. The parameter r can be replaced by an exponential function of transport costs, if a discrete choice framework for aggregate agent is used as an underlying theoretical basis for deriving the gravity model.

The only case known to us of an empirical test of this model is that of the Netherlands, where Davydenko & Tavasszy (2013) estimated these models separately for the three categories of O/D tables, using flow matrices obtained from statistics that denote the type of origin and destination. He finds plausible values for the parameters of the function determining the values of r and a good fit to observations.

4.3.2.3   Explicit Choice Models

A discrete choice modelling approach is applied to explain the step from P/C to O/D tables using a logic which is similar to the one developed in micro models, based on total logistics costs. In Figure 4.5, this model replaces the arrow between the two tables.

The choice to model is between alternative structures, either indirectly via an intermediate point of inventory, or directly between P and C. Note that for indirect flows, in principle, every region can be an intermediate point of inventory. In practice, these options via different regions will form one group of alternatives under the heading 'indirect transport', leading to a nested model form (Figure 4.6).

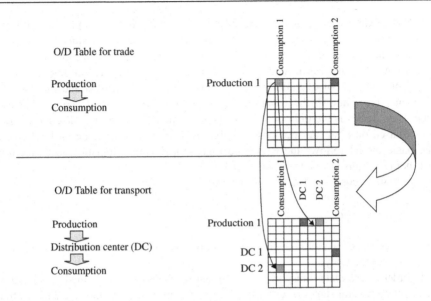

Figure 4.5 P/C table and O/D table (example: one DC each flow).

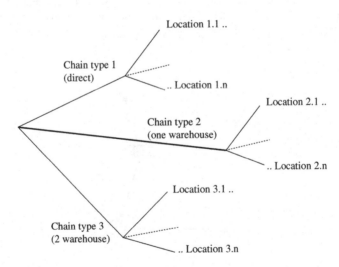

Figure 4.6 Nested model form for alternative distribution structures.

The few examples in the research literature that present such a model propose a logit formulation for the choice model (Tavasszy, Huijsman, & Cornelissen (2001), Jin & Williams (2005), Kim, Park, Kim, & Lee (2010), Davydenko & Tavasszy (2013) and two propose an all-or-nothing model (Boerkamps & van Binsbergen

(1999) & Maurer (2008). Kim et al. (2010) are the only ones who present a disaggregate choice model, however not based on cost logic, but on product characteristics. The drawback of this approach is that the model is less useful for policy evaluations. Tavasszy et al. and Jin and Williams develop aggregate logistic costs model but little proof of validity of the model has been provided and the models were only presented in the research literature to a limited extent. To our knowledge, Davydenko is the only published result on the estimation of a discrete choice model based on total logistics costs, at the level of observations, i.e. at O/D level. The basic model follows the aggregate logit model with a logistics cost function that includes transport and warehousing costs.

The logistic cost c has three components:

$$c = c^t + c^v + c^m$$

with superscripts t, v and m denoting transport costs, inventory costs and warehousing costs, respectively.

In order to calculate transport costs, alternative chains are subdivided in segments representing several stages of every chain type. Figure 4.7 shows the segments within a chain. The logic behind the segments is that the value of time (VOT), the modes available and the transport tariffs differ by stage. For example, segment S4 represents the final leg of a chain between a DC and a customer. As the need to deliver the goods on time is relatively high, it will have higher transport costs (usually this part is delivered with LTL (Less than Truck Load) trucks or vans) and a higher VOT.

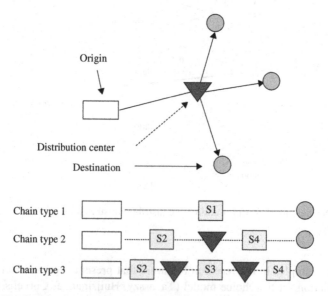

Figure 4.7 Distribution channels and segments (Tavasszy et al., 2001).

A simple weighted cost function (with weights according to the shares of different modes) was used to approximate transport costs. A condition limiting the calculation of weighted costs is provided by the modes that are allowed on different segments. On the last segment, S4, for example, only road transport is allowed. Inventory costs are specified for every freight flow and include the regular stock necessary for satisfying demand for a product and the safety stock. The warehousing costs are separated into handling and holding costs.

$$c^v = a_f X_{ij,f} i_f$$
$$c^m = a_f X_{ij,f} w(h_f + o_f)$$

i = interest costs (\$/tonne/year)
a_f = stock ratio per freight flow
$X_{ij,f}$ = trade flow on O/D-relation per flow type (tonne/year)
o_f = holding costs \$/m^3
h_f = handling costs \$/m^3
w = volume to weight ratio (m^3/tonne)

As in the micro modelling approach, the number of alternative locations for distribution centres, in theory, is equal to all available regions. For computational purposes, this number can be reduced by enumerating a limited number of alternative regions, which score well in terms of regional characteristics such as centrality with respect to demand. Procedures for this are described in Tavasszy et al. (2001).

Obviously, this aggregate model is a crude reproduction of the underlying aggregate cost patterns. If the data on costs are lacking, recourse must be taken to estimating the unknown cost elements as parameters in the models. Davydenko, Tavasszy & Smeets (2012) demonstrate that this approach can lead to a reasonably good fit of modelled with observed transport flows.

4.3.2.4 Supernetworks

The third alternative approach is to integrate the problem of choice of distribution centre into the route choice problem. To our knowledge, no empirical applications have been attempted yet with this approach for the problem of choice of distribution structure. We will, however, briefly describe this option, as it is a straightforward and elegant way to treat the problem. The approach follows from the generic 'supernetwork' approach (Sheffi, 1992) and extends the multimodal route choice model used for freight transport (see e.g. de Jong, 2007; Jourquin & Beuthe, 1996; Tavasszy, 1998) to now include distribution centres as passing points with inventory costs. Proposals for model specifications in this direction were done by Nagurney, Ke, Cruz, Hancock, & Southworth (2002) and Yamada, Imai, Nakamura, & Taniguchi (2011). If we include networks of different shipment sizes before and after distribution centres, the choice of shipment size

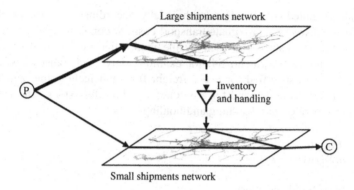

Figure 4.8 Supernetwork for inventory structures.

becomes implicit in the network.[1] Note that different techniques can be applied to predict flows on the network, depending on whether one assumes supply to be constant (demand-oriented choice models) or to be dependent of demand (equilibrium formulations).

In this network model, a switch from large to small shipment sizes, and low to high transport costs per unit freight, can only be made by using a distribution centre. Consumers will demand small shipment sizes and speedy delivery; direct delivery is then compared against indirect movement as alternative routing options (Figure 4.8).

The large shipment size networks are mainly intended for long distance trunk transport services. These types of shipments can be observed in international trade lanes such as Europe−Asia, US−Asia, where normally the smallest unit of shipment is a container. Large shipment size networks can also be observed at the level of countries or European Union. These are related to flows from production to the distribution facilities of retail chains, such as supermarkets. Freight aggregators and distributors also use these networks for inter-depot shipments, linking together large distribution facilities. Small shipment size networks, conversely, are linked to the goods flows to the consumption points or facilities functioning as a proxy to consumption, such as retail outlets. The small shipment size networks can also be observed in production environments too, such as specialized manufacturing and just in time production practices. These two networks are linked together through warehouses and other types of distribution facilities. Small shipment size networks serve the feeder function for the large shipment size networks: change from one network type to another happens at the distribution facilities.

---

[1] The inclusion of choice of shipment size also qualifies this problem as a hypernetwork problem (where alternative choices that are not made are represented by a network, while do not represent a physical alternative − the typical example for passenger transport would be time of day, or not to travel, see Sheffi (1992)).

As in the other two alternatives, the empirical application of the supernetwork approach relies on real-world observation of the logistics structures used by various sectors or types of businesses. There should also be observed flows within each of the two network types, which are not switched, such as small shipment networks between production and consumption (observed in case if production and consumption are located nearby, or in case of highly valuable goods). The large shipment network is also used for direct production—consumption shipments: these are mainly related to large batches of intermediate goods that are further used for production purposes.

4.4 Conclusion

Distribution structures are a relatively new area of concern in freight modelling. Including these structures enriches the conventional four-step freight model scheme, as they improve the representation of spatial flow patterns, of logistics costs and of intermediate storage activities. The earliest models for distribution structures were proposed around the turn of the century, and most of the empirical research is still under development. After applications in a handful of countries, these models are slowly entering the arena of applied national and international freight models.

This section has provided an overall description of modelling approaches at the disaggregate and aggregate level. Important first findings from research include (1) that micro-level normative models can also be applied to describe the result of firm level distribution decisions and (2) simple aggregate discrete choice models can achieve this for distribution structures in an interregional setting. The simplified disaggregate models shown here have the benefit of being close to firm level practice and thus sensitive to changes in firm or commodity attributes. At the same time, an important drawback appears to be that they require too much data to allow application at an aggregate level, for all freight flows. The reverse holds for aggregate models. With a relatively low demand for data, total flows can be modelled at a higher spatial and sectorial level. Note that even these aggregate models may require data acquisition efforts beyond what is currently undertaken in most modelling environments, including flows through warehouses and logistics cost components.

Future research can focus on bringing the micro- and macro level perspectives closer together. On the one hand, micro-level models could be developed that can deal with input data that are sufficiently aggregate to allow system wide acquisition, i.e. across all sectors and all regions. On the other hand, aggregate models can be better detailed, to provide a representative account for those complex considerations in distribution logistics that can now only be accounted for single firms or small groups of firms. In substantive terms, this includes empirical knowledge about the distribution structure decisions, their inclusion in micro or sector models, and their further aggregation, in order to arrive at better regional or national models for distribution structures.

References

Aikens, C. H. (1985). Facility location models for distribution planning. *European Journal of Operational Research, 22,* 263–279.

Bartholdi, J., & Hackman, S. (2008). Warehouse and distribution science, Release 0.85. Free Textbook, 2008. – URL <http://www.warehouse-science.com> Accessed 19.01.09.

Baumol, W. J., & Wolfe, P. (1958). A warehouse-location problem. *Operations Research, 6* (2), 253.

Beasley, J. E. (1988). An algorithm for solving large capacitated warehouse location problems. *European Journal of Operational Research, 33,* 314–325.

Beuthe, M., Vandaele, E., & Witlox F. (2004). Total logistics cost and quality attributes of freight transportation. In: *Selected papers of the WCTR 2004* in Istanbul.

Blauwens, G., De Baere, P., & Van de Voorde, E. (2012). *Transport Economics.* Berchem: De Boeck. ISBN: 978 90 455 4144 0.

Boerkamps, J., & van Binsbergen, A. (1999). GoodTrip – a new approach for modelling and evaluation of urban goods distribution. In: *Urban transport conference, 2nd KFB Research conference,* Lund, Sweden.

Burmeister, A. (2000). Familles logistiques: propositions pour une typologie des produits transportés pour analyser les évolutions en matière d'organisation des transports et de la logistique. Lille: PREDIT 'Systèmes d'information', 2000. Convention DRAST no 98 MT 87.

Cooper, L. (1972). The transportation-location problem. *Operations Research, 20,* 94–108.

Davydenko, I., & Tavasszy, L. A. (2013). Estimation of warehouse throughput in a freight transport demand model for the Netherlands. In: *Proceedings of the ninety second annual TRB meeting.* Washington, DC: Transportation Research Board.

Davydenko, I., Tavasszy, L. A., & Smeets, P. S. G. M. (2012). Commodity freight and trip generation by logistics distribution centers based on sectorial employment data. In: *Proceedings of the ninety first annual TRB meeting.* Washington, DC: Transportation Research Board.

de Jong, G., & Ben-Akiva, M. (2007). A micro-simulation model of shipment size and transport chain choice. *Transportation Research B, 41*(9), 950–965.

Domschke, W. (1981). *Logistik–transport.* München: Oldenbourg Verlag.

Domschke, W., & Drexl, A. (1996). *Logistik standorte* (4th ed.). München: Oldenbourg Verlag.

Friedrich, H. (2010). *Simulation of logistics in food retailing for freight transportation analysis* (Dissertation). KIT, Institut für Wirtschaftspolitik und Wirtschaftsforschung (IWW), published online: <http://digbib.ubka.uni-karlsruhe.de/volltexte/1000020602>.

Geoffrion, A. M. (1979). Making better use of optimization capability in distribution. *IIE Transactions, 11,* 96–108.

Green, G., Kim, C. S., & Lee, S. M. (1981). A multicriteria warehouse location model. *International Journal of Physical Distribution and Logistics Management, 11,* 5–13.

Jin, Y., Williams, I., & Shahkarami, M. (2005). Integrated regional economic and freight logistics modelling: results from a model for the Trans-Pennine Corridor, UK. In: *European transport conference.*

Jourquin, B., & Beuthe, M. (1996). Transportation policy analysis with a geographic information system: the virtual network of freight transportation in Europe. *Transportation Research Part C, 4*(6), 359–371.

Kim, E., Park, D., Kim, C., & Lee, J. -Y. (2010). A new paradigm for freight demand modeling: the physical distribution channel choice approach. *International Journal of Urban Sciences, 14*(3), 240–253.

Liedtke, G. (2009). Principles of micro-behavior commodity transport modelling. *Transportation Research E, 45*, 795–809.

Maurer, H. (2008). *Development of an integrated model for estimating emis-sions from freight transport* (Ph.D. dissertation). The University of Leeds Institute for Transport Studies.

Miehle, W. (1958). Link-length minimization in networks. *Operations Research, 6*, 232–243.

Nagurney, A., Ke, K., Cruz, J., Hancock, K., & Southworth, F. (2002). Dynamics of supply chains: a multilevel (logistical–informational–financial) network perspective. *Environment and Planning B: Planning and Design, 29*, 795–818.

Neumann, K., & Morlock, M. (1993). *Operations research*. München, Wien: Carl Hanser Verlag. ISBN 3-446-15771-9.

Ortúzar, J. de D., & Willumsen, L. (2011). *Modelling transport*. West Sussex: John Wiley & Sons.

Park, J.-K. (1995). *Railroad marketing support system based on the freight choice model* (Dissertation). Department of Civil and Environmental Engineering, M.I.T.

Pawelleck, G. (1996). Simulationsgestützte distributionsplanung. In: *Zeitschrift für Logistik, 17*, 6–9.

Sharma, R. R. K., & Berry, V. (2007). Developing new formulations and relaxations of single stage capacitated warehouse location problem (SSCWLP): empirical investigation for assessing relative strengths and computational effort. *European Journal of Operational Research, 177*, 803–812.

Sheffi, Y. (1992). *Urban transportation networks*. Englewood Cliffs, NJ: Prentice-Hall.

Späth, H. (1975). *Cluster-analyse-algorithmen*. München: Oldenbourg Verlag.

Tavasszy, L. A., Huijsman E., & Cornelissen C. (2001). Forecasting the impacts of changing patterns of physical distribution on freight transport flows in Europe. In: *Proceedings of the nineth world conference for transport research*.

Tavasszy, L. A. (2000). On the demand-side segmentation of freight transport markets with logistical families, THINK-UP Project Deliverable, Brussels: European Commission.

Tavasszy, L. A., Smeenk, B., & Ruijgrok, C. J. (1998). A DSS for modelling logistics chains in freight transport systems analysis. *International Transactions in Operational Research, 50*(6), 447–459.

Tempelmeier, H. (1980). *Standortoptimierung in der marketing-logistik*. Würzburg: Physica-Verlag.

Toporowski, W. (1996). *Logistik im Handel, Optimale Lagerstruktur und Bestellpolitik einer Filialunternehmung* (Dissertation). Universität Köln.

Yamada, T., Imai, K., Nakamura, T., & Taniguchi, E. (2011). A supply chain-transport supernetwork equilibrium model with the behaviour of freight carriers. *Transportation Research E, 47*(6), 887–907.

5 Inventory Theory and Freight Transport Modelling

François Combes

Université Paris-Est, LVMT, UMR T9403 ENPC IFSTTAR UMLV, Marne-La-Vallée, Cedex, France

Freight is transported by batches called shipments. Determining shipment size is an important decision of shippers; it derives from the characteristics of the transport supply and also from the logistical context of shippers.

Optimal shipment size is originally a question of inventory theory, related to the optimal management of a supply chain. However, it is also related to many transport-related decisions, in a way which sheds an instructive light on the preferences of shippers. The objective of the chapter is to present how models from inventory theory, originally derived from a normative perspective, can be used to better understand and model freight transport from a descriptive perspective. Freight mode choice is the main field benefiting from concepts of inventory theory, but other issues of freight transport modelling are also raised. The perspective of the chapter is that of microeconomics and econometrics; microsimulation, which also borrows elements from inventory theory, is let aside.

The chapter proceeds as follows: in Section 5.1, elements of inventory theory relevant to the topic of freight transport modelling are reminded. In Section 5.2, microeconomic models of shippers based on inventory theory are presented. Section 5.3 discusses the topic of data needs and availability. The current relevant econometric results are summarised in Section 5.4. Finally, Section 5.5 presents the challenges related to simulation when the choice of shipment size is taken into account, either explicitly or implicitly.

5.1 Inventory Theory

Inventory theory is a field of operations research focused on optimising the decisions of a firm operating a supply chain. It covers such themes as production, supply chain network design and transport decisions. For models of inventory theory related to freight transport, transport decisions are merely a facet of the larger problem of managing a supply chain. The objective, for a model of inventory theory, is

Modelling Freight Transport. DOI: http://dx.doi.org/10.1016/B978-0-12-410400-6.00005-7

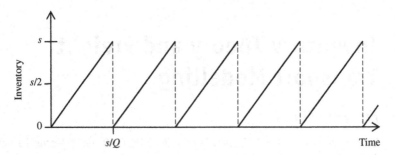

Figure 5.1 Evolution of the origin inventory in the EOQ model.

generally to optimise a certain objective function consisting of logistic costs, quality of service (or level of service) for customers and transport costs. This section presents the principles of a number of models of inventory theory which are particularly interesting from the perspective of freight mode choice modelling.

5.1.1 The Economic Order Quantity Model

The economic order quantity (EOQ) model was designed a century ago (Harris, 1913). It gives the optimal size of a production batch in a factory. In the context of this model, large batches allow for cost savings because setting up the machines for the production of a given batch is costly, independently of the batch's size. However, large batches cause increased logistic costs, because they have to be stocked. Consequently, there is an optimal batch size, balancing these two costs.

Translating the original EOQ model to the context of freight transport is straightforward (Larson, 1988). Consider a shipper sending a commodity flow rate Q (in tons per year, for example) from origin O to destination D. The shipper has to decide the size s of shipments. If the commodity is produced at a constant rate, then the inventory at the origin increases at rate Q at the origin from zero at time zero to s at time Q/s, at which point the first shipment is dispatched and the inventory drops back to zero, and then increases again at rate Q and so on. The evolution of the inventory level at the origin is illustrated by Figure 5.1. Clearly, its expected value is $s/2$.

A large inventory is something the shipper would like to avoid; this is represented in the EOQ model by a coefficient a, called the value of time of the shipper, which represents the amount the shipper is willing to pay to have one ton of commodity spend one day less in the supply chain (i.e., in this case, in the origin inventory). This coefficient is multiplied by the average inventory level to obtain the origin[1] inventory cost $as/2$, which increases with s. This is typically referred to as the 'cycle inventory' cost. To this component should be added the 'pipeline inventory' cost, related to the

---

[1] In some instances, it is relevant to also take into account the destination inventory costs. This is the case if the shipper is still the owner of the commodities at the destination or if, for some reason, they internalise the logistical costs borne by their customers. This is not the option taken in this paper, and it has little consequences − either theoretical or econometrical − for freight mode choice modelling.

commodities which are being transported at a certain time. Denote by t the transport lead time (i.e. the span between the moment the commodities are picked up and the moment they are delivered); the average amount of commodities being transported at a given moment is Qt, the corresponding cost is aQt.

On the contrary, too frequent shipments are also undesired, because they would incur unreasonably high transport costs. At this stage, it is necessary to make the relationship between shipment size and transport costs explicit. It is also necessary to depart from a classic hypothesis in freight transport modelling, according to which transport costs can be expressed on a per ton basis. Let us note $p(s)$ the cost[2] of transporting a shipment of size s from O to D. In the EOQ model, p is assumed to have the following form: $p(s) = b + cs$, where both b and c are in monetary units. In general, p does not necessarily have this particular specification. However, it is necessary that p does not tend towards zero when s tends towards zero; otherwise, it would be feasible for shippers to send infinitesimal shipments at an indefinitely high frequency.

The transport costs over a given time period are equal to $p(s)$ multiplied by the shipment frequency s/Q. By adding the inventory costs to the transport costs, we obtain the total logistic cost (TLC) function g:

$$g(s) = \frac{as}{2} + aQt + \frac{Q}{s}p(s) \tag{5.1}$$

which becomes, after replacing p:

$$g(s) = \frac{as}{2} + \frac{Qb}{s} + Q(c + at) \tag{5.2}$$

The optimal shipment size s^* is obtained by differentiating g and finding its minimum; this yields the classic square root EOQ formula:

$$s^* = \sqrt{\frac{2Qb}{a}} \tag{5.3}$$

The shipment size increases with the shipment size independent component of the transport cost b and decreases with the value of time of the shipper a. It also increases with the total commodity flow rate Q, although not linearly. And finally, it does not depend on the shipment size dependent component of the transport cost c.

When the shipment size is optimal, the cost function becomes:

$$g(s^*) = \sqrt{2Qba} + Q(c + at) \tag{5.4}$$

---

[2] This cost comprises the freight rates, the other monetary costs and the non-monetary costs associated to transporting a shipment of size s, from the perspective of the shipper.

5.1.2   Extensions of the EOQ Model

The EOQ model has been extended to a very large and varied set of contexts of applications. A few of them are listed below.

5.1.2.1   The Capacity Constraint of Vehicles

Accounting for the limited capacity of vehicles in the EOQ model is straightforward (Hall, 1985). Denote by K the capacity constraint of the vehicle; then the optimal shipment size becomes:

$$s^* = \min\left(\sqrt{\frac{2Qb}{a}}, K \right) \tag{5.5}$$

The cost function becomes:

$$g(s^*) = \begin{cases} \sqrt{2Qab} + Q(c + at) & Q \le \dfrac{K^2 a}{2b} \\[2ex] \dfrac{aK}{2} + \dfrac{Qb}{K} + Q(c + at) & Q > \dfrac{K^2 a}{2b} \end{cases} \tag{5.6}$$

This function has two regimes (Figure 5.2): when Q is low, the optimal shipment size is given by Eq. (5.3), and the cost function behaves as in the no capacity constraint case. When Q exceeds this threshold, shipments are of size K, and the cost function becomes linear.

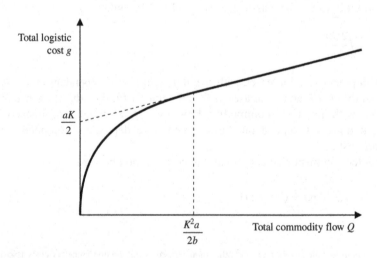

Figure 5.2 Influence of the commodity flow rate on the TLC.

5.1.2.2 One-to-Many Distribution: Rounds

One-to-many (or many-to-many) distribution refers to the case where commodities have to be carried from one (or many) origin(s) to many destinations, and where it is potentially advantageous to carry them together. This is a vast issue in operations research, closely related to the vehicle routing problem (VRP). The VRP is mathematically complex, demanding in terms of data, and sensitive to its initial conditions.

However, it is possible to address this problem with a statistical, approximate approach, which may have potential regarding freight transport modelling. This is illustrated below with an example inspired by Daganzo (2005).

Consider an area of size A, with n possible destinations. At each destination, deliveries are randomly requested. The arrival process of delivery requests at a destination is a Poisson process of parameter q (during a time interval t, qt deliveries are requested on average at each destination). As a consequence, the number of deliveries requested on the area per period of time is $Q = qn$ on average. Now, assume the shipper sends the deliveries from a unique warehouse somewhere in the area. Vehicles are dispatched by groups of n_v vehicles, every H hours, to deliver the commodities. Figure 5.3 illustrates these assumptions.

With these assumptions, the average value of the inventory at the origin is $QH/2$. Denote by a the value of time of the shipper; let $\alpha_c = aQ/2$. Then the per hour cycle inventory cost is:

$$C_c(H) = \alpha_c H \tag{5.7}$$

Now, in order to proceed, it is necessary to know how the transportation cost depends on H. At each time period, the n_v vehicles have to deliver QH destinations. As explained in Daganzo (2005), provided n_v is small w.r.t. QH, the expected distance $d(H)$ between two consecutive stops in an optimal tour can be approximated by $d(H) = 3/4(A/QH)^{1/2}$. As a consequence, given the fact that each vehicle has to

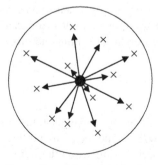
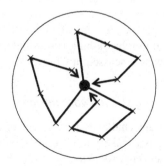

Commodity origins and destinations   Vehicle movement

Figure 5.3 One-to-many distribution.

deliver QH/n_v stops in a time interval H, and that each vehicle's speed[3] is v, then, necessarily, $(QH/n_v)d(H)/v = H$. This yields the required number of vehicles $n_v(H)$. Let c denote the operation cost of a vehicle for an hour. Then the per hour total operation cost is $C_o(H) = cn_v(H)$ or, with $\alpha_o = 3c(AQ)^{1/2}/4v$:

$$C_o(H) = \frac{\alpha_o}{\sqrt{H}} \tag{5.8}$$

Finally, if each vehicle finishes its tour in H hours, and if the deliveries are regularly spread along the route, then the average transport time for each commodity is $H/2$. The per hour pipeline inventory C_p cost is then (with $\alpha_p = aQ/2$):

$$C_p(H) = \alpha_p H \tag{5.9}$$

The TLC function g is the sum of Eqs. (5.7)–(5.9):

$$g(H) = \frac{\alpha_o}{\sqrt{H}} + (\alpha_c + \alpha_p)H \tag{5.10}$$

Consequently, the optimal headway is:

$$H^* = \left(\frac{3c\sqrt{A}}{8vaQ}\right)^{2/3} \tag{5.11}$$

This result is similar to the EOQ model (where $H^* = s^*/Q$, i.e. $H^* = (2b/aQ)^{1/2}$). The higher the transport operation cost c, but also the area's size A (for a given total level of demand), the lower the shipment frequency. On the contrary, the higher the vehicle speed v (for a given operation cost), the higher the shipper value of time, or the higher the total demand Q (i.e. either more frequent delivery requests q, or more potential destinations n), the higher the vehicle dispatch frequency.

Daganzo (2005) presents many similar models. These models may be a fruitful direction to improve freight transport demand modelling when tours are involved (for example in the context of urban freight transport, and also for transport services involving break-bulk operations). Urban freight modelling and urban logistics are otherwise discussed in Chapter 9.

5.1.2.3 Logistic Network Design

An important challenge in the current efforts to improve freight transport demand modelling is to take into account logistic networks. The location and throughput of warehouses, cross-docking platforms have a decisive impact on the freight transport

---

[3] Strictly speaking, the average speed v depends on $d(H)$: the lower the distance between two consecutive stops, the lower is v. This is disregarded here for analytical simplicity.

system, and adapt to it; yet they are difficult to observe and model. This problem is discussed in detail in Chapter 6.

In this situation again, models inspired from the EOQ model may provide insights into why certain logistic network structures are preferred to other. Consider for example the case of a factory serving a set of destinations, all located far from the factory but close to one another. Then it may be useful to install a distribution centre near the destinations, carry the commodities from the factory to the distribution centre with big, fast vehicles, then dispatching them to the individual destinations in smaller quantities, to keep both transport costs and inventory costs at a reasonable level. This, however, would come at the cost of renting and operating the platform, and transhipping the commodities. Simple analytical models can be used to compare such options.

5.1.2.4 Other Extensions

As stated above, this category of models is quite large and varied. Some of them focus on the interaction between production constraints and transport constraints (Blumenfeld, Burns, Diltz, & Daganzo, 1985). Some of them generalise the EOQ model to various freight price structures (Aucamp, 1982; Langley, 1980). Others base an optimal vehicle size model on an underlying EOQ shipment size model (McCann, 2001). Daganzo (2005) presents many extensions of the continuous approximation method originally presented in Newell (1971), which is a simple method to extend the models presented above to cases where the parameters are not uniform, but vary slowly (e.g. a slow variation of Q in time, or a slow variation of the density of destinations in space). All these models are possible resources to extend freight transport modelling. However, the benefit they may bring in terms of behavioural relevance should be balanced with the additional data needs they would come with.

5.1.3 Models of Optimal Shipment Size with Uncertain Demand

Logistics is about delivering goods or providing services both at a low cost and a high level of service. This does not only involve finding the right trade-off between transport costs, inventory costs and customer satisfaction but also involves doing so in an uncertain context. In this section, a simple model of choice of shipment size with stochastic demand is briefly presented.

5.1.3.1 The Single Commodity Periodic Review Model with Backlogging

The daily management of a supply chain under an uncertain context is a very active field of operations research and management science. Often, the objective is to provide firms with efficient algorithms to operate a supply chain under stochastic conditions.

The single commodity periodic review model with backlogging is a classic model of inventory theory. It is presented in Arrow, Karlin, & Scarf (1958). In this model, a shipper produces a certain type of commodity at an origin O, and sends them at a given moments, say each day, to a destination D where they are sold to consumers. Those consumers present themselves randomly, without notice, to buy the commodities; they wait if there is a momentary stock shortage, but it causes them dissatisfaction. Each day, the shipper has to decide the size of the shipments which are dispatched. However, transport takes a certain time l from O to D, so that the shipper has to anticipate the demand at destination. Each day, the destination inventory increases by the size of the shipment sent from the origin l days before, and decreases by the amount ordered by customers during the day (it can get negative if there are orders pending).

Assume that the demands D_t at time periods t are independent and identically distributed, with a common cumulative distribution function F_D. The optimal policy for the shipper is to ensure that the destination inventory is on average equal to a certain value I_s, the safety stock. The adequate value of I_s balances the expected destination inventory costs (which increase with I_s) and the expected customer satisfaction (which decreases when stock shortages become more probable and longer).

Assume the demands are normally distributed, of mean μ and standard deviation σ; denote by Φ the Gaussian c.d.f. Let a be the customers' willingness to pay to avoid waiting a one day delay, a_c the commodity value of time for the shipper (e.g. capital opportunity costs and depreciation costs) and a_w the destination warehousing costs, all in monetary units per ton per day. Then, the optimal safety stock minimising the expected TLC of the shipper is:

$$I_s^* = \sigma\sqrt{l+1}\,\Phi^{-1}\left(\frac{a}{a_c + a_w + a}\right) \tag{5.12}$$

The safety stock increases (in absolute value) with the variability of the final demand; the more difficult it is to predict the behaviour of consumers, the more the shipper will have to protect itself against unpredictable variations. Also, and perhaps more importantly from the perspective of freight transport modelling, the safety stock increases (again, in absolute value) when the transport lead time increases. This indicates that a slower transport mode is not beneficial to the shipper's logistics. Finally, the sign and amplitude of the safety stock depends on the relative values of the logistic costs a, a_c and a_w. In fact, if $a > a_c + a_w$, then the safety stock is positive, and it increases with a: the more customers are sensitive to shortages, the higher the safety stock. In the symmetric case, the shipper opts for a smaller safety stock. It may even be optimal to aim for a negative safety stock, in other words to have the customers wait almost certainly.

5.1.3.2   Implications

At this stage, the applications of these findings are mainly operational. They are devised to guide firms in operational decisions, not to provide models of shipper behaviour on the freight transport market. Besides, variables such as safety stock, as shipment size, may appear not to be of direct interest for freight transport modelling. However, the microeconomic analysis of the models presented in this section yields particularly instructive results on the behaviour of shippers, and on the relationship between these variables and freight transport-related decisions, and particularly freight mode choice.

Note that the EOQ model addresses the problem of the optimal shipment frequency, while the inventory replenishment model addresses that of optimal safety stock. A model bridging the gap between these two partial models exists: the classic (s, S) logistic protocol (a shipment of size $S - s$ is dispatched as soon as the destination inventory level decreases below s). This protocol is widely used, makes both the shipment size and the safety stock variables endogenous, and is mathematically optimal (Porteus, 2002). However, in order to be useful for freight transport modelling, an analysis of the statistical properties of this model is necessary, and it has not been undertaken yet.

5.2   Microeconomics of Logistics and Freight Transport

Inventory theory adopts the point of view of a single shipper, adapting to a given economic context, and aims at providing an optimal decision rule. Freight transport modelling adopts the point of view of many shippers considered together, and focuses on their reactions to changes in the economic context, given the decision rules they do apply. The objective of this section is to proceed to this change of perspective: the objective is not anymore to devise the optimal behaviour of a shipper in a given situation. It is to model the behaviour of shippers on the freight transport market, assuming they follow decision rules derived from inventory.

This section proceeds in three stages. First, the general notion of TLC is introduced. Then, a simple model of freight mode choice based on the EOQ model is presented. Extensions of this model are finally discussed.

5.2.1   The TLC Function

Inventory theory generally requires that a certain objective function be optimised. For example, in the case of the EOQ model, the optimal shipment size minimises the function g given by Eq. (5.2). In the case of the one-to-many distribution model, the optimal headway minimises the function g given by Eq. (5.10). The safety stock I_s^* given by Eq. (5.12) also optimises a certain function (not given here).

These functions share some similarities with the generalised cost functions classically met in freight transport modelling. Generalised cost functions represent the preferences of agents between distinct transport alternatives. They generally consist in monetary components and non-monetary components, weighted by the relevant marginal substitution rates. There is a neo-classical theoretical underpinning of this approach for passenger transport (Fowkes, 2010; Jara-Díaz, 2007), which is in fact an application of theories of the allocation of time (Becker, 1965). In freight transport, the theory is less clear (although some elements are presented in the previous section), but the practice is the same.

The difference between generalised cost functions and the functions met in Section 5.1 is that the latter include logistic cost components: origin or destination inventory costs, cycle inventory and safety inventory costs, sometimes also ordering costs, handling costs, customer satisfaction, etc. Therefore, they are referred to as TLC functions.

In theory, a model based on TLC functions offers a much broader spectre of possibilities regarding decision support. Aside from being sensitive to classic variables such as travel times and costs, it also reacts to changes in warehousing costs, handling costs, vehicle capacity constraints, customer preferences, etc. This is one of the main reasons why much effort has been put into trying to integrate inventory theory considerations in freight transport models.

5.2.2   The TLC in the EOQ Model: A Simple Freight Mode Choice Model with Logistics

The EOQ model, in addition to yielding an optimal shipment size, associates to each mode a certain TLC. Denote by m a certain transport mode, with a travel time t_m, a transport cost $p_m(s) = b_m + c_m s$ and a capacity K_m. Then, Eq. (5.6) is the TLC associated to choosing mode m:

$$g_m = \begin{cases} \sqrt{2Qab_m} + Q(c_m + at_m) & Q \leq \dfrac{K_m^2 a}{2b_m} \\[4mm] \dfrac{aK_m}{2} + \dfrac{Qb_m}{K_m} + Q(c_m + at_m) & Q > \dfrac{K_m^2 a}{2b_m} \end{cases} \tag{5.13}$$

Clearly, a freight mode choice model can be built with these TLC functions. Now, to illustrate the potential of such a model, consider the choice between two modes, a light one (indexed by l) and a heavy one (indexed by h). Assume that road is faster ($t_l < t_h$), less expensive for small shipments ($b_l < b_h$) but more expensive on a per ton basis ($c_l > c_h$), and with vehicles of a lower capacity ($K_l < K_h$). Then, depending on the values of a and Q, the shipper will prefer the road if $g_l < g_h$ (Section 5.2.4 comes back on the assumptions related to transport costs).

In classic freight transport demand models, the generalised cost is proportional to the commodity flow rate Q so that it is not necessary to keep track of Q as an explanatory variable of mode choice. This is not the case anymore with this approach inspired from inventory theory. The evolution of the TLC of the two modes with Q is illustrated by Figure 5.4:

The total commodity flow has a dramatic impact on the relative competitiveness of each mode. The light mode is more competitive for small flows. Those imply small shipments, which are less expensive to transport when b is smaller. However,

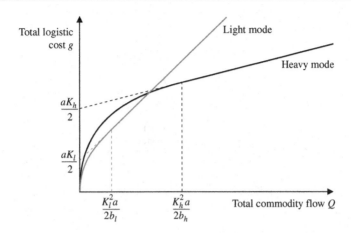

Figure 5.4 Comparison of the TLCs of two modes.

c is larger for the light mode, and its vehicle capacity K is lower, so that when the commodity flow increases, the light mode quickly loses its advantage. Symmetrically, the heavy mode is too costly for small commodity flows: setting up a transport operation is costly and inefficient for small shipments. But, for large flows, the vehicle's increased capacity and lower c prove decisive.

It is also interesting to examine the domains where the two modes are competitive, depending on the shipper–receiver characteristics a and Q. For the case without the capacity constraint (Eq. 5.4), the light mode is preferred to the heavy mode when a is larger than a certain threshold $\bar{a}(Q)$ given by:

$$\bar{a}(Q) = \frac{c_l - c_h}{t_h - t_l} \left(1 - \frac{2}{1 - \sqrt{1 + 4(t_h - t_l)(c_l - c_h)Q/\left(\sqrt{b_h} - \sqrt{b_l}\right)^2}} \right) \quad (5.14)$$

Figure 5.5 shows the domains of competitiveness of the two modes in the (a, Q) plane. Consistently with Figure 5.4, for a given commodity value of time, the heavy mode is preferred for larger commodity flows. Conversely, for a given commodity flow, the light mode is preferred for larger values of time. Note that for large values of Q, the asymptotical value of time threshold is familiar: it is the one obtained from classic freight mode choice models. These calculations are more complex when the capacity constraints are accounted for; however, the general model behaviour remains. The empirical relevance of this model, and particularly of Figure 5.5, is discussed in Section 5.4.2.

This example shows how simply inventory theory can be applied to mode choice modelling. It shows that the important perspective for freight mode choice is that of the shipper–receiver relationship, and that the shipper–receiver commodity flow rate plays an essential role. This is confirmed empirically in Section 5.4.

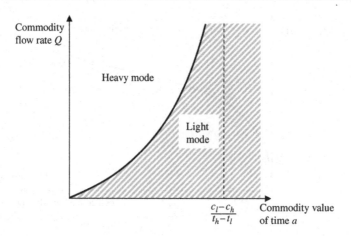

Figure 5.5 TLC functions and mode competitiveness.

5.2.3 The TLC in the Context of a Dynamic Model: A Partial Theory of the Value of Time and Value of Reliability in Freight Transport

The inventory replenishment model presented in Section 5.1.3 gives instructive results when analysed from a microeconomic perspective. How do the logistic and transport costs of the shipper behave when the transport lead time l increases? It is possible to obtain a closed-form expression of the shipper TLC's derivative with respect to l and thus of the shipper's willingness to pay for travel time savings (Combes, 2010). Denote by α this value of time:

$$\alpha = a_c + \frac{\zeta}{2\sqrt{l+1}} \frac{\sigma}{\mu} \tag{5.15}$$

where ζ is a parameter increasing with a, a_c and a_w. The value of time consists of two components. The first one corresponds to the pipeline inventory cost: if the transport time decreases, then the commodities spend less time in the transport operation, which is beneficial. The second part is related to the destination logistic cost. It mirrors the fact that a faster transport mode allows for a better control of the supply chain. The shipper can reduce the destination safety stock as well as the probability and expected duration of shortages, thus improving the satisfaction of their customers.[4] Note that the shipper is all the more sensitive to travel time as the demand at destination is variable, if customers are highly sensitive to shortages, or if destination warehousing costs are large.

---

[4] Even though customer dissatisfaction is not a cost borne by shippers, they should take it into account (Mohring, 1985). Indeed, if a shipper can, to a cost of less than a euro, improve the customer's satisfaction for a euro's worth, they should do so and increase the commodity's price by one euro. The customer will be indifferent, and the shipper will have improved their margin.

From a microeconomic perspective, this result has significant implications. First, a given shipper may have different values of time for different modes. Second, improvements to travel time are worth more for already fast travel times. Third, the usual econometric assumption that the willingness to pay for travel time savings is an exogenous characteristic of the shipper does not hold.

Along this modelling approach, it is also possible to calculate the willingness to pay for an improvement in travel time reliability. Denote by σ_l the variability of travel time. Then, the value of reliability is[5] (Combes, 2009, p. 255−256)

$$\alpha_r = \frac{\zeta}{\sqrt{1 + (l + 1)\sigma^2/\mu^2\sigma_l^2}} \tag{5.16}$$

and the value of time becomes:

$$\alpha = a_c + \frac{\zeta}{2\sqrt{(l + 1) + \mu^2\sigma_l^2/\sigma^2}}\frac{\sigma}{\mu} \tag{5.17}$$

These results give a theoretical ground to the value of reliability in freight transport: unreliable travel times are bad because they cause destination inventory variability, and thus higher logistic costs and customer dissatisfaction. Also, it shows that there may be a non-linear relationship between the value of time and the value of reliability. In particular, if the reliability of a particular mode is poor, the associated value of time will be reduced; improvements in speed are less useful for the users of an unreliable transport mode.

Again, there are important econometric implications to Eq. (5.16). Indeed, the value of reliability is very small when the travel time reliability is high. Therefore, the estimation of the value of reliability of shippers using a reliable mode will unsurprisingly be low. It would still be a mistake to conclude that shippers do not pay attention to reliability.

To conclude, it should be reminded that the relevance of the inventory replenishment model is limited by the fact that shipment frequency is exogenous. The work which was undertaken for this model should now be led with the more general (s, S) model described in Porteus (2002), to check whether the conclusions above hold.

5.2.4 The Inventory Theoretic Model of Freight Transport Demand of W. J. Baumol and H. D. Vinod

Currently, the most general specification of a TLC function for a freight mode choice model is given by Baumol & Vinod (1970). In their model, Baumol and Vinod consider the same cost components as in Eq. (5.2), to which they add

---

[5] These results currently involve a conjecture; more theoretical work is still needed to completely assert their validity.

another cost component related to the safety stock. This stock is assumed to be equal to $k\sqrt{(H + l)}\mu$ where k is chosen so that there is only a given probability that a stock shortage happens, and H is the average headway between two shipments.

The TLC function in Baumol & Vinod (1970) is (after a few adaptations):

$$g(H) = \frac{aQH}{2} + \frac{b}{H} + Q(c + at) + (a_c + a_w)k\sqrt{(H + l)}\mu \qquad (5.18)$$

The cost resulting from the uncertainty of the demand is the additional warehousing and commodity cost due to the safety stock; the customer cost due to stock shortages is ignored. With this model, it is neither possible to obtain a closed-form expression for the optimal shipment frequency, nor for the TLC associated to each mode.

This model has two limitations: first, when the demand is uncertain, the optimal logistic protocol is the (s, S) one, not a fixed shipment frequency one. Second, the model does not account for the customer costs, and k is exogenous. However, it includes both the logic of the EOQ model and that of the inventory replenishment model; and its theoretical limitations may have only a limited empirical impact. Therefore, it is a potentially excellent basis for an econometric model of freight mode choice based on inventory theory. It has inspired many of the empirical models of freight mode choice discussed in Section 5.4.

5.2.5   The Structure of Freight Transport Costs

As already discussed when presenting the EOQ model, the structure of freight transport costs should be paid special attention. A transport mode cannot be characterised by a per ton cost alone; more accurate models of transport costs are required.

First, the optimal shipment size depends directly on the shape of the transport costs. Obviously, different transport cost structures yield different optimal sizes (Langley, 1980; Vernimmen & Witlox, 2003), and also different TLC functions. However, there are few papers actually studying the relationships between shipment size and freight rates. Swenseth & Godfrey (1996) present a typical curve of rates as function of shipment size; Smith, Campbell, & Mundy (2007) estimate the relationship between freight rates, shipment size, annual volume, and distance; Kay & Warsing (2009), focus on less-than-truckload (LTL) freight rates in the United States.

Furthermore, it is interesting to look at the reasons behind the structure of transport costs. This is in fact related to the production functions of carriers, which are far from trivial when shipment sizes are taken into account. Let us distinguish the truckload (TL) and the LTL segments. TL transport is not very efficient for pickup operations but is organised so as to ensure high loading factors, and thus limited per ton transport costs. On the contrary, LTL transport is efficient for pickup and delivery, but overall more expensive on a per ton basis. With the notations of Section 5.1.1, $b_{TL} > b_{LTL}$ and $c_{TL} < c_{LTL}$. Given the results of Section 5.2.2, classic transport will be preferred for larger flows, with bigger shipments, whereas consolidation transport will be preferred for smaller flows, with smaller shipments. For a more detailed discussion of the relationship between carriers' production costs and rates in freight transport, see Combes (2013).

5.3 Databases

In order to assess empirically models inspired by inventory theory, adequate databases are necessary. This section presents two surveys or groups of surveys which are particularly relevant to our topic: the commodity flow surveys (CFSs) in Section 5.3.1, and the French ECHO (Envois Chargeurs Opérateurs) survey in Section 5.3.2. Note that a more general discussion of freight transport data is available in Chapter 11.

5.3.1 Commodity Flow Surveys

CFSs are shipper surveys, where the observation unit is the shipment. Each surveyed shipper has to describe the nature and transport characteristics of a sample of shipments dispatched during a given time period. CFSs typically contain millions of observations.

Let us consider the example of the US CFS of 2007. It covers the business establishments located in the United States and belong to a given set of production and service activities. 102,000 establishments were surveyed, each of them reporting information on 20–40 shipments. It defines a shipment as: '*a single consignment of commodities or products from [the] establishment to a single customer or to another specific location of [the] company transported in commerce, often with a shipping document such as a manifest, bill of lading or waybill*'. It makes clear that if some shipments are transported together on part of their trips (e.g. during a round), they should still be considered as distinct shipments (US Department of Transportation and US Department of Commerce, 2010).

Each shipper declares the total number of outbound shipments during the surveyed week. Then, for a sample of shipments, information about the shipment date, value, net weight, commodity type is collected. The destination and the transport modes are observed. If the shipment is placed in a piece of transportation equipment designed to be interchanged between different modes, the transport is considered intermodal. If the shipment is exported, the exit gateway and the transport mode to the gateway are observed.

Another example of CFS is the Swedish one. Its most recent edition was in 2004/2005, and it is similar to the US CFS. About 12,000 establishments were surveyed, for a total of approximately 749,000 shipments (Swedish Institute for Transport and Communication Analysis, 2006).

CFSs are useful for the estimation of freight mode choice models based on inventory theory. However, they observe the decisions of shippers, but they do not observe some of the variables underlying the decisions of shippers, such as the shipper—receiver commodity flow rate. This is an important limitation when trying to estimate models inspired by inventory theory.

5.3.2 The ECHO Shipper Survey

The French shipper survey ECHO, realized in 2004/2005, is similar to a CFS. It is a shipper survey, collecting information on shippers and shipments. However, it differs by the range of information it includes. In particular, it observes the shipper—receiver relationships and the shipment transport operations in a very detailed way (Guilbault, 2008; Guilbault & Gouvernal, 2010). Up to 700 variables are observed for each shipment; however, only 10,462 shipments are observed.

The information in the ECHO database can be categorised into four groups:

1. Establishment: The shipper and receiver are described by their locations, activities, volumes of commodities sent or received, access to infrastructures or equipments, production and logistic processes, etc.
2. Shipment: Characteristics of the shipment (size, weight, nature, value, conditioning, etc.), characteristics of the shipper—receiver relationship (flow rate, communication means, deadlines, etc.)
3. Transport operation (physical chain): Origin and destination, number of legs, characteristics of each leg, transhipment location, packaging, transport time and cost, etc.
4. Transport operation (organisational chain): Decision-making levels, number of operators, service performed, subcontracting relationship, etc.

The ECHO survey is particularly interesting from the perspective in inventory theory. It is the unique database to observe essential variables such as the shipper—receiver commodity flow rate.

5.3.3 Inventory Theory and the Need for Adequate Data

Taking into account shipment size in freight models comes with specific data needs. Concerning transport demand, data availability is not excellent, but some databases exist. They are not sufficient, at this stage, to easily build full-fledged spatialised simulation models, but they allow for research to progress. Some variables such as safety stocks, demand variability, travel time reliability, potentially play an important role, and are not observed yet.

On the side of transport supply, the situation is less favourable. There is little information on the relationship between shipment size and the transport performances of various modes, and particularly prices. The topic of the relationship between freight rates and shipment sizes is not simple, but cannot be ignored. Freight rates mirror three things: the cost of the carriers' inputs (drivers, vehicles, fuel, etc.), the carriers' market power, and how well the transport of a particular shipment fits in the operation of a carrier (see Demirel et al., 2010 on the last point). This makes the estimation of freight rates a complex empirical issue. Fortunately, the absence of reliable data on freight rates does not entirely prevent the estimation of mode choice models based on inventory theory.

5.4 The Econometrics of Freight Mode Choice and Shipment Size

Most of the empirical analyses of freight transport involving concepts of inventory theory, or shipment size, concern freight mode choice. In this section, the choice was made to categorise them into two groups: exploratory analyses (Section 5.4.1), where the authors test specifications inspired by general microeconomic considerations, intuitions, and the variables available; and structural analyses, where inventory theory is explicitly referred to by the authors as a motivation for the choice of specification (Section 5.4.2). It should be noted that however different these postures may be, they sometimes lead to very similar specifications.

5.4.1 Exploratory Analyses

From an econometric perspective, the relationship between freight mode choice and shipment size has been identified a long time ago. It began with studies highlighting the causal influence of freight mode choice on shipment size, and conversely, and then proceeded with studies focused on the interdependency between both decisions.

5.4.1.1 Causal Models

A number of studies undertaken from the late 60s to the early 80s show that freight mode choice depends on shipment size. Miklius (1969) estimates a mode choice model where shipment size is an explanatory variable. The estimation shows that bigger shipments increase the probability that rail be the transport mode. Conversely, Rakowski (1976) shows that shipment size depends strongly on mode choice. Friedlander & Spady (1980) also use shipment characteristics as explanatory variables of mode choice, in the frame of a classical approach involving an aggregate cost function. To each mode is associated a sort of generalised price which depends, among other variables, on the average characteristics of shipments. Winston (1981) reaches similar results by estimating a multinomial probit model of mode choice: the probability to be carried by rail is increased for shipments bigger than a certain threshold. All these papers are based on various editions of the US CFS. In a later work, Jiang, Johnson, & Calzada (1999) also used shipment characteristics as explanatory variables of mode choice, based on the French *Enquête Chargeurs* of 1988, the ancestor of the French ECHO survey.

5.4.1.2 Simultaneous Models

Other estimations show that both variables are interrelated. They consist in estimating discrete—continuous models of mode and shipment size. The estimation of a simultaneous model is not trivial, and a number of different methods are applied. Abdelwahab & Sargious (1992), and Abdelwahab (1998), based on the US CFS of 1977, consider the choice between rail and truck. Mode choice is assumed to be ruled by a binomial probit model. The latent variable is denoted I^* and specified as (with minimal adaptations):

$$I_i^* = \gamma X_i + \eta_{\text{truck}} Y_{\text{truck},i} + \eta_{\text{rail}} Y_{\text{rail},i} + \varepsilon_i \tag{5.19}$$

where X_i is a vector of exogenous variables. $Y_{\text{truck},i}$ and $Y_{\text{rail},i}$ are the shipment sizes conditionally to the choice of one mode (they are observed for the chosen mode only):

$$\begin{aligned} Y_{\text{truck},i} &= \beta_{\text{truck}} X_{\text{truck},i} + \varepsilon_{\text{truck},i} \quad \textit{if f} \quad I_i^* > 0 \\ Y_{\text{rail},i} &= \beta_{\text{rail},i} X_{\text{rail},i} + \varepsilon_{\text{rail},i} \quad \textit{if f} \quad I_i^* \leq 0 \end{aligned} \tag{5.20}$$

The two models are estimated simultaneously, and the estimates of the correlations between ε, $\varepsilon_{\text{truck}}$ and $\varepsilon_{\text{rail}}$ are significant, thus showing that the decisions of mode choice and shipment size are empirically interrelated.

The second approach, presented in Holguín-Veras (2002), is a bit more structural. Using a specific data set collected in Guatemala City, a model of shipment size is first estimated independently of vehicle choice. Then, the choice between vehicle types is specified as a discrete choice model based on, for each mode m, the variable XL_m specified as follows:

$$XL_m = \left| M_m - y \right| \tag{5.21}$$

where y is the shipment size predicted by a specific model, and M_m an indicator of the payload for mode m. M_m can be defined in different ways, some of them leading to a better statistical fit than others.

The third approach relies on more recent statistical methods. Pourabdollahi, Karimi, & Mohammadian (2013) couple a multinomial logit mode choice model and a multinomial logit shipment size with a Copula distribution (regarding Copula distributions, refer e.g. to Bhat & Eluru, 2009). The estimation also shows the interdependency between both decisions. The model is estimated using a specific shipper database (Sturm, Pourabdollahi & Mohammadian, 2013) collected by the University of Illinois, Chicago, in order to support the development of the model freight activity microsimulation estimator (FAME), an activity-based model for freight transport (Samimi, Pourabdollahi, Mohammadian & Kawamura, 2012).

5.4.2 Structural Analyses

In this section, different types of models are presented: models of shipment size, models of mode choice, and models of simultaneous shipment size and mode choice. They all build on the theoretical elements presented in Sections 5.1 and 5.2.

5.4.2.1 Choice of Shipment Size

The theory of optimal shipment size is not recent. The estimation of a model of optimal shipment size is, in principle, very easy. Indeed, consider the optimal shipment size given by Eq. (5.3). It is equivalent to:

$$\ln s^* = \frac{1}{2}\ln 2 + \frac{1}{2}\ln Q + \frac{1}{2}\ln b - \frac{1}{2}\ln a \tag{5.22}$$

an equation which is easily assessed empirically, provided the relevant database is available. The model was estimated with the ECHO database in Combes (2012). The commodity value density a_{dens} is taken as a proxy of the commodity value of time a, and the total commodity flow rate Q_{tot} is taken as a proxy of the commodity flow rate Q. Freight rates are not reliable in the ECHO database, so that b was simply considered as a mode-specific constant. The following equation was estimated:

$$\ln s^* = \beta_m + \beta_Q \ln Q_{\text{tot}} + \beta_a \ln a_{\text{dens}} + \varepsilon \tag{5.23}$$

where m is the transport mode. The estimates are very consistent with the theoretical EOQ model. The estimate of β_Q is close to 0.5, and that of β_a close to -0.5; the hierarchy of the β_m is consistent with the discussion in Section 5.2.2: higher β_m go with heavier modes (on one hand, for short to medium transport: $\beta_{\text{waterway}} > \beta_{\text{rail}} > \beta_{\text{combined}} > \beta_{\text{private carrier}} > \beta_{\text{common carrier}}$, and on the other hand, for overseas transport: $\beta_{\text{sea}} > \beta_{\text{air}}$). The estimation also shows that, in the case of road transport, more transhipments in the transport operation are associated to smaller shipments, an observation which confirms that direct transport is suitable for large commodity flows (and thus large shipments), while consolidation is relevant for smaller flows. On the whole, the estimation of the EOQ model explains about 80% of the variance in the ECHO database.

5.4.2.2 Mode Choice

As discussed in Section 5.2.4, when applying the EOQ model, the TLC associated has a certain specification. A mode choice model can then be designed with such TLC functions. This is done by Lloret-Batlle & Combes (2013) with the ECHO database. The explained variable is then the mode choice, and the explanatory variables those suggested by inventory theory, such as a_{dens} and Q_{tot}. From Eq. (5.6), a theoretical per ton TLC function is formulated.

$$\frac{g_m}{Q} = \begin{cases} \sqrt{\dfrac{2ab_m}{Q}} + c_m + at_m & Q \leq \dfrac{K_m^2 a}{2b_m} \\[3mm] \dfrac{aK_m}{2Q} + \dfrac{b_m}{K_m} + c_m + at_m & Q > \dfrac{K_m^2 a}{2b_m} \end{cases} \tag{5.24}$$

In Lloret-Batlle & Combes (2013), the following specification (and variants) is used:

$$V_m = ASC_m + \beta_{\text{EOQ},m}\sqrt{\frac{a_{\text{dens}}}{Q_{\text{tot}}}} + \beta_{d,m}d + \beta_{t,m}a_{\text{dens}}d + \varepsilon_m \qquad (5.25)$$

Based on these utilities functions, mode choice is given by a multinomial logit model. For each mode, the corresponding utility depends on an alternative specific constant ASC_m, the ratio $\sqrt{a_{\text{dens}}/Q_{\text{tot}}}$, the distance d and the product of the distance d and the value density a_{dens}. The objective of the two latter variables is to capture respectively the transport cost and the time of each mode, which are not available in the ECHO database.

According to theory and also to the estimation of the EOQ model, the coefficient $\beta_{\text{EOQ},m}$ should be lower for heavier modes. This is confirmed by the estimation of this model: the coefficients of private carrier and common carrier are indistinct; those of combined transport, inland waterway and rail transport are significantly lower. For overseas transport, the coefficient of sea transport is lower than that of air transport. At this stage, Lloret-Batlle & Combes (2013) is more a proof of concept. It illustrates the important role of a_{dens} and Q_{tot}, but it lacks important variables to be directly useful as a freight mode choice model.

Figure 5.5 also confirms qualitatively the theory underlying Eqs. (5.24) and (5.25). It represents the main transport mode for each shipment in the ECHO database. The eight sub-graphs correspond to eight distance bands, ranging from 0 to 20,000 km. In each subgraph, the x-axis is a_{dens}, the value density of the commodity (in €/t), while the y-axis is Q_{tot}, the total commodity flow rate from the shipper to the receiver (in t/y), both in logarithm scales. Each dot corresponds to a shipment, the symbol representing the main transport mode (generally the non-road one in a multimodal transport chain, see Guilbault, 2008, for the exact definition of the main mode in the ECHO database).

Figure 5.6 is consistent with theoretical obtained Figure 5.5. Heavier modes are preferred for larger Q_{tot} and lower a_{dens}. This is particularly visible in the case of the competition between road and rail for distances between 150 and 1000 km; and between sea and air transport for distances larger than 3000 km.

5.4.2.3 Joint Mode Choice and Shipment Size

An alternative approach is to model mode choice and shipment size simultaneously. Its advantages are that it uses the observed shipment size for the estimation, and that it accounts very simply for the vehicle capacity constraints: shipment size cannot exceed the capacity constraint of a vehicle, because it is simply not an alternative in the model. Its downside is that the estimation process is not simple.

Two main category of models can be distinguished. First, the discrete–discrete choice models, where shippers can choose among a finite (but potentially large) set of (mode combination, shipment size) couples. The advantage of this modelling approach is that the classic discrete choice specifications can be readily applied.

Vertical axis log (Q_{lot}) horizontal axis log (a_{dens})

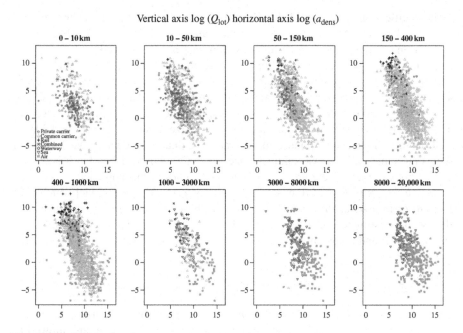

Figure 5.6 Commodity flow rate and value density, origin−destination distance, and mode choice in the ECHO database.
Reproduced with permission from: Combes, F., Ruijgrok, K., & Tavasszy, L. (2013). Endogenous shipment size in freight mode choice models: theory and empirical testing. In M. Ben-Akiva, H. Meersman, & E. Van de Voorde (Eds.), *Freight transport modelling.* Bingley, UK: Emerald Group Publishing Limited.

Each (mode, shipment size) couple is characterised by a deterministic utility function, inspired by TLC functions such as that of Baumol and Vinod (Eq. 5.16).

In the frame of a larger project discussed in Section 5.5, De Jong & Ben-Akiva (2007) estimate a multinomial logit discrete−discrete model with 5 shipment size categories and 17 mode combinations. To each (mode, shipment size) couple is associated a utility function of which the specification is mostly determined by the availability of relevant data. For example, the annual shipper−receiver commodity flow rate is unobserved, and therefore not taken into account. The explanatory variables kept for the estimation include the access to an industrial rail track or to a quay at the origin, the company size, the commodity type, the value density of the shipment, the transport cost and the product of the shipment value and transport duration. Attention was paid to reproduce precisely the transport costs: link-based costs and transhipment costs are taken into account. The model was estimated using the Swedish CFS of 2001. More flexible specifications (nested logit and mixed logit) did not lead to better results.

Windisch, de Jong, van Nes, & Hoogendoorn (2010) estimated a similar model on the Swedish CFS 2004/2005, but 14 shipment size categories and 8 transport chains were distinguished; some couples were forbidden. Multinomial and nested logit specifications were tested. Habibi (2010) also estimated a similar model with the same database. All these studies met the limitation of the Swedish CFS, i.e. the fact that the shipper−receiver commodity flow rate is not observed.

Abate (2012) estimates a model of truck size choice and choice of shipment size, with a Danish heavy vehicle trip diary data collected in 2006 and 2007. The explanatory variables of the truck size model include the total volume of freight expedited by shippers, without distinction of receivers, the age of vehicles, fleet size, etc. Then, the shipment size model also includes truck size as an explanatory variable.

The second category relates to discrete−continuous models. This approach seems more natural, because shipment size is a continuous variable, but it is technically more complex. There are currently few examples of this approach.

McFadden, Winston, & Boersch-Supan (1985) present a joint model which is in fact closer to a sequential model. It is loosely based on principles from inventory theory. It is estimated on US transport data of 1977. The explained variables are shipment size and mode choice (between truck and rail). Shipment size is specified as follows:

$$s = \beta_1 t_{truck} + \beta_2 t_{rail} + \beta_3 b_{truck} + \beta_4 b_{rail} + \beta_5 c_{truck} + \beta_6 c_{rail} + \varepsilon_1 \qquad (5.26)$$

where t_m is the travel time of mode m, b_m (resp. c_m) the shipment size independent (resp. dependent) component of the freight rates of mode m (which are allowed to depend on the distance), and ε a normally distributed error term. The mode choice itself is modelled by a latent variable F:

$$F = \gamma_0 + \gamma_1 (t_{truck} - t_{rail}) + \gamma_2 (b_{truck} - b_{rail}) + \gamma_3 (c_{truck} - c_{rail}) + \gamma_4 a_{dens} + \gamma_5 s + \varepsilon_2$$
$$(5.27)$$

where the value density of the commodity a_{dens} is an explanatory variable of mode choice. The error term ε_2 is also normally distributed, and possibly correlated with ε_1. Truck is chosen if $F > 0$.

De Jong & Johnson (2009) do remind the principles of inventory theory before presenting a continuous−discrete model of mode choice and shipment size. The specification of the model itself is similar to that of Holguín-Veras (2002), presented in Section 5.4.1. Its performance is compared to other model architectures. One of the conclusions is that the estimates can vary significantly with the specification. Another conclusion is that in principle the model should be estimated using full information maximum likelihood. If not possible, two-step estimation methods can be applied; however, this requires that the explanatory variables do not appear both in the shipment size equation and in the mode choice equation. In that case, it is not possible to have both shipment size and mode choice sensitive to e.g. transport costs, which is not satisfying.

5.4.2.4 Implications

Three general conclusions can be drawn from this wide array of empirical studies. First, it is interesting to note that despite the fact that the primary objective of inventory theory is to provide normative models, it is now also a fruitful toolset to understand and model freight transport from a descriptive standpoint. Second, the commodity flow rate from the shipper to the receiver is an essential explanatory variable of mode choice, as shown by the estimation of inventory theoretic models against the ECHO database. Its absence from current models and databases is a serious obstacle to the accuracy of freight traffic studies. Third, other variables, such as safety stocks, final demand variability, etc., may also have a potentially very important impact on freight mode choice, and yet are not observed at all. As a direct consequence, survey protocols should be extended to account to the variables, such as shipper−receiver commodity flows, which play such an important role in the preferences and decisions of shippers.

These methodologies have not been integrated in freight modelling practice yet, but it is an actively investigated field, and it should lead to operational mode choice models in a potentially short time.

5.5   Perspectives for Simulation

Integrating principles of inventory theory in freight transport models is a complex task. The theoretical and empirical discussions above have shown that this involves specific theoretical and observation challenges.

The first difficulty is of theoretical nature. Models of inventory theory shed an interesting light on the behaviour of shippers, especially regarding the importance of the logistic context of freight transport. Accounting for variables such as safety stocks or customer demand variability would be an improvement both for traffic modelling and for socio-economic analyses. However, the relevant microeconomic framework is still unachieved.

The second difficulty relates to data availability. The example of the ECHO database is striking: by making a few strategic variables available, it has allowed to overcome a significant number of locks in the estimation of models from inventory theory. However, this database is unique, not recent, and of limited size. New surveys are needed, and survey protocols (such as the CFSs) need to be updated (maybe trading off the number of observations, which is high, with the number of variables).

The third difficulty is related to the architecture of models. For example, one clear conclusion from this chapter is that the relevant modelling framework is the shipper−receiver pair. Instead of the classic approach, where the demand for freight transport on a given origin−destination pair is described by the aggregate commodity flows per commodity type, it is necessary to describe it by a joint distribution of (Q, a) describing the population of shipper−receiver pairs. The mode choice stage should operate at the disaggregate level of those pairs. Then, once the commodity flows are allocated to the various transport modes available between each origin−destination pair, it is necessary to translate them into vehicle movements. This, again, is not trivial: while it is reasonable to assume that TL shipments fill the vehicles and thus that one shipment is equal to one vehicle, this is not the same anymore for LTL shipments, where consolidation is most often involved (this line of reasoning clearly applies to all transport modes). Consolidation involves tours, break-bulk platforms, in other words economies of scales, which raise a considerable amount of issues (this is discussed in Chapter 8).

Most of these principles were already presented in De Jong, Ben-Akiva, Florian, Grønland, & Van de Voort (2005). The authors also presented the architecture of a model developed along these lines: the aggregate—disaggregate—aggregate (ADA) model. The objective of this model, also discussed in De Jong, Ben-Akiva, & Baak (2007), De Jong & Ben-Akiva (2007) and Ben-Akiva & de Jong (2013), is to convert aggregate production—consumption (PC) matrices into vehicle flows, while modelling the choice of logistic chain,[6] shipment size, mode choice and transport operation in line with principles from inventory theory. The first stage is to disaggregate the PC flows to obtain a synthetic population of shippers and receivers, to assign them to one another, and then to assign the commodity flows to the resulting shipper—receiver pairs. Then, a joint discrete—discrete mode choice and shipment size model is applied. Finally, shipments are aggregated and converted into vehicle movements. During the last step, the consolidation of shipments is handled explicitly. The estimation of the mode choice/shipment size model is discussed in Section (5.4.2); the remaining parameters are calibrated so that the observed aggregate commodity flows are reproduced. At this stage, this is one of the most ambitious freight models regarding the introduction of inventory theory principles.

References

Abate, M. (2012). The optimal truck size choice. Presented at Kuhmo-Nectar conference, Berlin, Germany.

Abdelwahab, W. M. (1998). Elasticities of mode choice probabilities and market elasticities of demand: evidence from a simultaneous mode choice/shipment size freight transport model. *Transportation Research Part E, 34*(4), 257—266.

Abdelwahab, W. M., & Sargious, M. A. (1992). Modeling the demand for freight transport: new approach. *Journal of Transport Economics and Policy, 26*(1), 49—72.

Arrow, K. J., Karlin, S., & Scarf, H. (1958). *Studies in the mathematical theory of inventory and production*. Stanford, CA: Stanford University Press.

Aucamp, D. C. (1982). Nonlinear freight costs in the EOQ model. *European Journal of Operational Research, 9*, 61—63.

Baumol, W. J., & Vinod, H. D. (1970). An inventory theoretic model of freight transport demand. *Management Science, 16*(7), 413—421.

Becker, G. (1965). A theory of the allocation of time. *Economic Journal, 75*, 493—517.

Ben-Akiva, M., & de Jong, G. (2013). The aggregate—disaggregate—aggregate (ADA) freight model system. In M. Ben-Akiva, H. Meersman, & E. Van de Voorde (Eds.), *Freight transport modelling*. Bingley, UK: Emerald Group Publishing Limited.

---

[6] An otherwise actively investigated field of freight transport modelling, discussed in Chapter 6.

Bhat, C., & Eluru, N. (2009). A copula-based approach to accommodate residential self-selection effects in travel behaviour modelling. *Transportation Research Part B, 43*, 749−765.

Blumenfeld, D. E., Burns, L. D., Diltz, J. D., & Daganzo, C. (1985). Analyzing trade-offs between transportation, inventory and production costs on freight networks. *Transportation Research Part B: Methodological, 19*(5), 361−380.

Combes, F. (2009). *The choice of shipment size in freight transport* (Ph.D. Thesis). Université Paris-Est, Marne-la-Vallée.

Combes, F. (2010). Logistic imperatives and mode choice. In: *Proceedings of the European transport conference*, Glasgow, United Kingdom.

Combes, F. (2012). Empirical evaluation of the economic order quantity model for choice of shipment size in freight transport. *Transportation Research Record: Journal of the Transportation Research Board, 2269*, 92−98.

Combes, F. (2013). On shipment size and freight tariffs: technical constraints and equilibrium prices. *Journal of Transport Economics and Policy, 47*(2), 229−243.

Combes, F., Ruijgrok, K., & Tavasszy, L. (2013). Endogenous shipment size in freight mode choice models: theory and empirical testing. In M. Ben-Akiva, H. Meersman, & E. Van de Voorde (Eds.), *Freight transport modelling*. Bingley, UK: Emerald Group Publishing Limited.

Daganzo, C. (2005). *Logistics systems analysis* (4th ed.). Springer.

De Jong, G., & Ben-Akiva, M. (2007). A micro-simulation model of shipment size and transport chain choice. *Transportation Research Part B, 41*, 950−965.

De Jong G., Ben-Akiva, M., & Baak, J. (2007). A micro-model for logistics decisions in Norway and Sweden calibrated to aggregate data. In: *Proceedings of the European transport conference*. Leiden, Netherlands.

De Jong, G., Ben-Akiva, M., Florian, M., Grønland, S. E., & Van de Voort, M. (2005). Specification of a logistics model for Norway and Sweden. In: *Proceedings of the European transport conference*. Strasbourg, France.

De Jong, G., & Johnson, D. (2009). Discrete mode and discrete or continuous shipment size choice in freight transport in Sweden. In: *Proceedings of the European transport conference*. Leiden, Netherlands.

Demirel, E., van Ommeren, J., & Rietveld, P. (2010). A matching problem for the backhaul problem. *Transportation Research Part B, 44*, 549−561.

Fowkes, A. S. (2010). The value of travel time savings. In E. Van de Voorde, & T. Vanelslander (Eds.), *Applied transport economics − a management and policy perspective*. Antwerpen: De Boeck.

Friedlander, A. F., & Spady, R. H. (1980). A derived demand function for freight transportation. *The Review of Economics and Statistics, 62*(3), 432−441.

Guilbault, M. (2008). Enquête ECHO: Envois-Chargeurs-Opérations de transport: résultats de référence, INRETS.

Guilbault, M., & Gouvernal, E. (2010). Transport and logistics demand: new input from large shippers surveys in France. *Transportation Research Record: Journal of the Transportation Research Board, 2168*, 71−77, Transportation Research Board of the National Academies, Washington, DC.

Habibi, S. (2010). *A discrete choice model of transport chain and shipment size on the Swedish Commodity Flow Survey 2004/2005* (Masters thesis). Kunglika Tekniska Högskolan.

Hall, R. W. (1985). Dependence between shipment size and mode in freight transportation. *Transportation Science, 19*(4), 436−444.

Harris, F. W. (1913). How many parts to make at once. *Factory, the Magazine of Management, 10*(2), 135–152.

Holguín-Veras, J. (2002). Revealed preference analysis of commercial vehicle choice process. *Journal of Transportation Engineering, 128*(4), 336–346.

Jara-Díaz, S. (2007). *Transport economic theory*. United Kingdom: Elsevier.

Jiang, F., Johnson, P., & Calzada, C. (1999). Freight demand characteristics and mode choice: an analysis of the results of modeling with disaggregate revealed preference data. *Journal of Transportation and Statistics, 2*(2), 149–158.

Kay, M. G., & Warsing, D. P. (2009). Estimating LTL rates using publicly available empirical data. *International Journal of Logistics Research and Applications, 12*(3), 165–193.

Langley, C. J. (1980). The inclusion of transportation costs in inventory models: some considerations. *Journal of Business Logistics, 2*(1), 106–125.

Larson, P. D. (1988). The economic transportation quantity. *Transportation Journal, 28*(2), 43–48.

Lloret-Batlle, R., & Combes, F. (2013). Estimation of an inventory-theoretic model of mode choice in freight transport. *Transportation Research Record: Journal of the Transportation Research Board*(in press).

McCann, P. (2001). A proof of the relationship between optimal vehicle size, haulage length and the structure of distance-transport costs. *Transportation Research Part A, 35*, 671–693.

McFadden, D., Winston, C., & Boersch-Supan, A. (1985). Joint estimation of freight transportation decisions under non-random sampling. In A. Daughety (Ed.), *Analytical studies in transport economics*. Cambridge, UK: Cambridge University Press.

Miklius, W. (1969). Estimating freight traffic of competing transportation modes: an application of the linear discriminant function. *Land Economics, 45*(2), 267–273.

Mohring, H. (1985). Profit maximization, cost minimization and pricing for congestion-prone facilities. *Logistics and Transportation Review, 21*, 27–36.

Newell, G. F. (1971). Dispatching policies for a transportation route. *Transportation Science, 5*(1), 91–95.

Porteus, E. L. (2002). *Foundations of stochastic inventory theory*. Stanford, CA: Stanford University Press.

Pourabdollahi, Z., Karimi, B., & Mohammadian, A. K. (2013). A joint model of freight mode and shipment size choice. In: *Proceedings of the 92nd annual meeting of the Transportation Research Board*, Washington, DC.

Rakowski, J. P. (1976). Competition between railroads and trucks. *Traffic Quarterly, 30*(2), 285–301.

Samimi, A., Pourabdollahi, Z., Mohammadian, A. K., & Kawamura, K. (2012). An activity-based freight mode choice microsimulation model. In: *Proceedings of the 91st annual meeting of the Transportation Research Board*. Washington, DC.

SIKA. (2006). *Commodity flow survey 2004/2005*, SIKA Report 2006:12, SIKA, Stockholm.

Smith, L. D., Campbell, J. F., & Mundy, R. (2007). Modeling net rates for expedited freight services. *Transportation Research Part E, 43*, 192–207.

Sturm K., Pourabdollahi, Z., & Mohammadian, A. K. (2013). A nationwide establishment and freight shipment survey: descriptive and non-response bias analyses. In: *Proceedings of the 92nd annual meeting of the Transportation Research Board*. Washington, DC.

Swenseth, S. R., & Godfrey, M. R. (1996). Estimating freight rates for logistics decisions. *Journal of Business Logistics, 17*(1), 213–231.

US Department of Transportation, & US Department of Commerce (2010). *2007 Economic census, transportation, 2007 commodity flow survey.*

Vernimmen, B., & Witlox, F. (2003). The inventory-theoretic approach to modal choice in freight transport: literature review and case study. *Brussels Economic Review, 46*(2), 5–28.

Windisch, E., de Jong, G. C., van Nes, R., & Hoogendoorn, S. P. (2010). A disaggregate freight transport model of transport chain and shipment size choice. In: *Proceedings of the European transport conference.* Glasgow, Scotland, UK.

Winston, C. (1981). A disaggregate model of the demand for intercity freight transportation. *Econometrica, 49*(4), 981–1006.

6 Mode Choice Models

Gerard de Jong

Institute for Transport Studies, University of Leeds, UK;
Significance BV, The Hague, The Netherlands; and Centre for Transport
Studies, VTI/KTH, Stockholm, Sweden

6.1 Introduction

6.1.1 Mode Choice at Different Spatial Levels

Mode choice or modal split models in freight transport explain the allocation of a given total freight transport demand in an area (or a given origin–destination (OD) matrix with total freight flows between origins and destinations) over the available transport modes. What the modes are depends on the spatial scale.

Most modal split models in freight transport have been developed for interurban (or interregional) transport flows (this includes most national freight models in Europe, state-level models in the United States, and even models for Europe or the United States as a whole). The modes from which one can choose for this spatial context usually include road and rail transport. Depending on the topography of the study area, inland waterway transport and short sea shipping can also be choice alternatives (for some of the OD pairs).[1] In a modal split model applied at the OD level, the choice set of available alternatives does not have to be the same for all OD combinations. Good practice in modal split modelling is to exclude unfeasible alternatives (such as road transport between an island and the mainland, or inland waterway transport for locations far away from an inland port) from the set of available alternatives.

For the urban context, the only available transport mode usually is road transport. Here it may make sense to distinguish between various types of road transport vehicles (sometimes called 'vehicle choice' as opposed to mode choice). Of course an interurban model might also distinguish several vehicle types within each mode, but for urban transport it is especially important to be able to distinguish between small and large road transport vehicles, since this has a large impact on the liveability in the city, and some truck sizes might even be banned from certain areas.

For intercontinental flows, the available modes are sea and air transport. In terms of tonnes transported, the share of air transport is very small, but it has a

---

[1] In principle also pipelines, but because the use of pipelines is so commodity- and location-specific this is hardly ever included in a mode choice model (more often pipeline transport is excluded beforehand).

Modelling Freight Transport. DOI: http://dx.doi.org/10.1016/B978-0-12-410400-6.00006-9

substantial share in terms of the value of the goods or the costs of the transport services.

6.1.2   Relevance of Modal Split

Modal split modelling is not only a vital component of the overall four-or-more-step freight transport model (in many situations it is impossible to understand the vehicle flows on the network links, without modelling mode choice), it is also an important explanatory factor for the emissions of freight transport. Sea and inland waterway transport have generally lower emission rates (though there often is considerable scope for improvements within these modes) per tonne-kilometre than rail transport, which in turn has a lower rate than road transport (air transport having the highest rate). Examples can be found in Maibach et al. (2008). But also the requirements on public funding and accident rates vary greatly between modes.

In many freight transport models, the modal split is the most policy-sensitive demand component, in the sense that it reacts more to changes in transport time and cost than transport generation and distribution (which in some models are not sensitive to policies at all). Network assignment/route choice on the other hand might be more sensitive to such changes. However, a discussion of the possible policy effects in freight transport should not be restricted to modal split, as sometimes happens. International and regional trade flows might be also be sensitive to changes in transport time and costs, and some of the choices that are often ignored in freight transport modelling, such as shipment size choice and the loading rates of the vehicles, can be influenced by transport time and cost. We will come back to these impacts of changes in transport time and costs when discussing the literature on elasticities in freight transport in Chapter 9. Some other choices in freight transport, which can be combined with mode choice, are mentioned below.

6.1.3   Dependent and Independent Variables

The dependent variable in mode choice can be a discrete choice (one of the modes is chosen the other modes are not chosen), when the analysis is carried out at the level of the individual shipment, or a fraction (modal share) within a certain geographic area of for a specific OD flow, when the analysis is carried out at a more aggregate level.

The explanatory variables for modal split can be transport cost of the available modes (including loading, and unloading cost), their transport time, (sometimes combined into generalised transport costs), the number of transshipments, reliability (referring to the degree of on time delivery), flexibility (ability to handle short-term requests), probability of damage during transport, tracking and tracing of the cargo, the harmful emissions and transport frequency offered. The sensitivity of the modal split to these attributes of the modes can for instance differ between different commodity types (e.g. bulk versus general cargo), shipment sizes, industries, firm sizes, transport equipment used (e.g. containers) and geographic distance.

6.1.4 Disaggregate and Aggregate Mode Choice Models

A key distinction in freight mode choice modelling is that between aggregate and disaggregate models. In fact in the context of aggregate models one often speaks about 'modal split' and within disaggregate models about 'mode choice', but in this book we use both terms as synonyms.

By 'disaggregate' we mean here that the unit of observation is the individual decision-maker (travellers in passenger transport; firms in freight transport) as opposed to 'aggregate' models where the units of observation are aggregates of decision-makers, usually geographical zones.

Disaggregate models are much less common in freight transport than in passenger transport. The main reason for this difference is the lack of publicly available disaggregate data on freight transport, which in turn is too a large extent due to the commercial nature of such data. Firms involved in freight transport are often reluctant to disclose information on individual shipments, mode chosen, transport cost, etc. In Chapter 10, we'll come back to the issue of data availability.

Within disaggregate freight transport models, the decision that has been modelled most is clearly the mode choice. For this reason, and because a single mode choice yields a relatively easy model, we present he disaggregate choice modelling theory as part of this chapter (Section 6.2), using mode choice as the relevant context. Different types of models for a single discrete choice (multinomial logit (MNL), probit, nested logit, ordered generalised extreme value, cross-nested logit, mixed logit, latent class) are discussed in this section, as well as deviations from utility maximisation.

However, the aggregate mode choice model, especially the aggregate logit model, is the most commonly used model for mode choice in freight transport. We think it is best to see this as a pragmatic approach, not as a model based indirectly on a theory of individual behaviour, that however regularly leads to satisfactory results (elasticities, forecasts) at relatively low effort (especially in data collection).

In Section 6.3 we discuss practical examples of both aggregate and disaggregate mode choice models.

6.1.5 Closely Related Choices

Mode choice is usually studied in isolation, i.e. as a single endogenous variable. Also in most regional, national or international freight transport forecasting systems the modal split is determined independently from the trade volumes and the level-of-service from the networks acts as one or more exogenous variables in mode choice. However, there is much to be said for freight model systems with multiple dependent variables that allow for simultaneous choice making on mode choice and other choices. Some other choices in freight transport are closely connected to that on the mode, and sometimes also modelled in a simultaneous fashion:

- *A series of mode choices in the form of a transport chain choice model.* Mode choice can be studied for an OD flow, which has the advantage that the choice alternatives can be simple and the choice set limited. However, it often happens in practice that the use of a specific mode is combined with the use of other modes in a transport chain. A transport chain is a sequence of modes and transshipment locations that are all used to transport shipments from the sender to the receiver (e.g. road—rail—road). So transport chains refer to the PC (production—consumption) level, not to the OD level. An example of a transport chain would be road—rail—road, which decomposes into two road OD flows and one rail OD flow. We think that modelling transport chain choice at the level of the PC flows is a qualitatively superior strategy to mode choice at the OD level, especially because the latter might lead to suboptimal solutions at the PC level. However, in practice, data might not be available for a transport chain choice model, since transport data are usually collected at the OD level. In a transport chain choice model, the choice alternatives are sequences of modes (including direct transport using a single mode all the way) instead of modes. A simplified way of modelling transport chain choice is to define a main mode at the PC level and model the choice between available main modes. This definition could be based on the longest distance or on a modal hierarchy (e.g. inland waterway is the main mode if it is used somewhere in the chain, otherwise rail is the main mode if it is used somewhere in the chain and if not, the main mode is road transport). The explanatory variables for inland waterways and rail could then include road transport costs of getting to and from the inland port or rail terminal and corresponding transshipment cost. The choice models in Section 6.2 can be regarded as OD mode choice models or PC main mode choice models.

- *Mode choice and shipment size choice model* (usually measured in tonnes). Smaller shipment sizes are almost always transported by road, and larger shipments have an increased probability of being transported by a non-road mode (e.g. rail, inland waterway transport). This by itself could be a reason to include shipment size as an exogenous variable in mode choice (e.g. Jiang, Johnson & Calzada, 1999). But one could go one step further and model two-way interactions where mode choice also influences shipment size choice. Holguín-Veras, Xu, de Jong, & Maurer (2011) carried out economic experiments with groups of students that were playing shippers and carriers, where the shippers knew the inventory costs function and the carriers knew the transport cost functions by mode. The carriers had to compete with each other to supply transport services to the shipper and did this by submitting a sealed bid (containing a price and a mode) to the shipper. This was repeated a number of times. These experiments rather soon converged to the joint optimum in terms of modes and shipment size, as predicted by game theory. The assumption that freight mode choice is an independent decision was not supported, and a joint mode and shipment size choice model is preferred. This leads to disaggregate models in which the mode choice decision is embedded in a larger inventory-theoretic and logistics framework, so that shipment size optimisation can be covered as well (more on this framework can be found in Chapter 5).
- *Models for the choice of mode and supplier* (in the sense of the origin zone for the transport flow). In passenger transport modelling one sometimes comes across joint models of mode and destination choice, where a traveller chooses between combinations of destination zones and modes to that destination, given the origin. In freight transport modelling, mode-destination choice does not seem a very sensible option, since it would amount to client choice by the suppliers. A more realistic option would be to take the viewpoint of the receiver of the goods (a firm that processes incoming goods, or a wholesaler or retailer) and model his choice of supplier jointly with the mode from that supplier (sender) to the given location of the receiver.
- *Mode and route choice models*. This includes disaggregate models that combine mode and route choice in freight transport in a simultaneous decision-making framework as

well as aggregate multi-modal network models. The multi-modal network modelling provides another way to handle transport chains.[2] In a transport chain, several modes are used consecutively for a door-to-door shipment. An example is to use a lorry first from the zone of the sender to the port, then use short sea shipping, then rail transport and finally lorry delivery to the zone of the receiver. Assignment to such combinations of modes in a transport chain can take place if the network not only includes links and nodes for each mode but also multi-modal nodes that connect one network to another network. Such nodes can be ports or rail and inland waterway terminals for transshipment between modes. In other words, a multi-modal network (or super network) is created, where intermodal transfer nodes for instance link road, rail and inland waterways networks.

Mode and shipment size choice are only modelled at the disaggregate level (that is for individual shipments). The other three joint choices can be modelled both at the aggregate and the disaggregate level, and especially mode and route choice is indeed mostly done at the aggregate level. Transport models for regional and national authorities and international organisations often have large networks that include so many mode and route choice alternatives (with difficult correlation structures, e.g. routes partly overlap), that a disaggregate joint mode and route choice model is usually not considered a feasible option (but this may change in the future). However, within a certain corridor, and especially for crossing a certain screenline (such as a sea strait or mountain range), there may only be a very limited number of route alternatives, and joint mode and route choice (both a discrete variables) is not cumbersome at all.

Examples of joint models (aggregate and disaggregate) of mode choice and related choices will be discussed in Section 6.3.

6.2 The Disaggregate Mode Choice Theory

6.2.1 Cost Functions and Utility Functions

As in most disaggregate mode choice models, we'll start by assuming that the decision-maker is the shipper, a firm that needs to send goods to a receiving firm and therefore has a demand for transport services. Shippers then have a choice to carry out the transport themselves or to contract it out to carriers (or more generally to logistics service providers, that also include firms that integrate services of different carriers for their customers). In practice, these firms in turn can also have a say in the mode choice. So, often in freight transport several firms (shipper, receiver, carriers and intermediaries) are involved in decision-making about the same shipment.[3] Instead of assuming that one of these firms takes the (mode choice) decisions, one could also try to model the interactions between the different parties involved, as joint decision-making. This is a relatively new area in freight

---

[2] The other way is to have a disaggregate or aggregate model for the choice from a choice set containing different uni-modal and multi-modal *transport chains* (as discussed above).

[3] Additionally, lorry drivers often have some freedom to choose the route, or adapt the route when facing congestion.

transport modelling. In Section 6.2.4, we discuss work carried out so far in this area.

The shipper that we assume decides, makes mode choice decisions for shipments. A shipment is defined as a number of units of a product that are ordered, transported and delivered at the same time. It doesn't necessarily correspond to a vehicle load, because there may be several (small) shipments in the same vehicle (consolidation of shipments), whereas a large shipment may require several vehicles.

The alternatives in mode choice can be road transport, rail transport, inland waterway transport, sea transport, air transport and pipeline transport. Furthermore, one could distinguish several vehicle or vessel types within these modes (vehicle type choice). An essential characteristic of all these alternatives is that these are discrete alternatives (as opposed to both continuous choice variables and ordered choice alternatives). The model for this is the discrete choice model.

Discrete choice models at the disaggregate level have originally been developed in passenger transport, where the dominant choice paradigm and theoretical foundation is that of random utility maximisation, RUM (McFadden, 1974, 1978, 1981). The basic equation of the RUM model is:

$$U_{ik} = Z_{ik} + \varepsilon_{ik} \tag{6.1}$$

In which:

U_{ik}: utility that decision-maker k derives from choice alternative i $(k = 1,\ldots K; i = 1,\ldots,I)$
Z_{ik}: observed utility component
ε_{ik}: random utility component.

Utility maximisation belongs to the economics of consumer behaviour, and seems at first sight inappropriate for explaining the behaviour of the firm, where the standard economic paradigm is that of profit maximisation or costs minimisation. However, we can apply the random utility framework to freight transport choices by simply using minus the total generalised transport costs[4] as the observed component of utility and including one or more random costs components to this function:

$$U_{ik} = -G_{ik} - e_{ik} \tag{6.2}$$

In which:

G_{ik}: observed component of generalised transport cost
e_{ik}: random cost component.

---

[4] The generalised transport costs are the direct monetary costs of transporting goods plus the influence of other qualitative characteristics of the modes (transport time, reliability, etc.) expressed in money units. In a model that also includes inventory considerations (such as the choice of shipment size) one could even generalise further and use total logistics costs (comprising amongst others transport and inventory costs).

In Eq. (6.2), random costs minimisation becomes RUM. For the standard discrete choice models with an independent error term and without heteroskedasticity, one might just as well write a $+$ sign in front of e_{ik}.The mode choice mode for freight transport can then be estimated using the same software as for the RUM model in passenger transport. One not only has to take into account that an increase in costs leads to a reduction in utility but also how transport cost works in the passenger transport model. If one assumes negative coefficient signs for the costs variables, one can also write $+ G_{ik}$ in Eq. (6.2), as we will do below.

More specifically, for the choice of mode for a specific shipment by decision-maker k, between three alternatives (road, rail and inland waterways (IWW)), one might specify the following linear[5] utility functions[6]:

$$U_{\text{road}} = \beta_0 + \beta_1 \cdot \text{COST}_{\text{road}} + \beta_2 \cdot \text{TIME}_{\text{road}} + \beta_3 \cdot \text{REL}_{\text{road}} + e_{\text{road}} \quad (6.3\text{a})$$

$$U_{\text{rail}} = \beta_4 + \beta_1 \cdot \text{COST}_{\text{rail}} + \beta_5 \cdot \text{TIME}_{\text{rail}} + \beta_6 \cdot \text{REL}_{\text{rail}} + e_{\text{rail}} \quad (6.3\text{b})$$

$$U_{\text{IWW}} = \beta_1 \cdot \text{COST}_{\text{IWW}} + \beta_7 \cdot \text{TIME}_{\text{IWW}} + \beta_8 \cdot \text{REL}_{\text{IWW}} + e_{\text{IWW}} \quad (6.3\text{c})$$

In which:

COST$_i$: transport cost of mode i; this could include both the distance-dependent costs $f_i \cdot$ DIST$_i$ (such as fuel costs), where f is the transport cost per km and DIST the distance in km, and the time-dependent cost $g_i \cdot$ TIME$_i$ (such as transport staff and vehicle cost), where g is the costs per hour.
TIME: transport time of mode i in hours.
REL$_i$: transport time reliability of mode i; this could be measured as the standard deviation of transport time or as the percentage of shipments delivered on time.
$\beta_0, \beta_1, \ldots, \beta_8$: coefficients to be estimated; we expect negative signs for $\beta_1, \ldots \beta_3$ and $\beta_5, \ldots \beta_8$, the sign for β_0 and β_4 can be positive or negative.

In Eqs. (6.3a)−(6.3c), the utility that would be obtained when choosing road transport depends on the transport cost and time for that shipment by road transport and its reliability, and likewise for the other two modes. The values for COST, TIME and REL by mode, may come from skimming networks for these modes, and also might be provided by the decision-makers themselves (often they find this hard for non-chosen modes, and there could be perception errors) or have been postulated in a 'what if' fashion by the researcher in a stated preference survey.

The βs are coefficients for which numerical values are determined by estimating the model on data for various decision-makers and corresponding individual

---

[5] Non-linear specifications of the utility function, such as functions with logarithmic or quadratic attributes, translog costs functions (e.g. Oum, 1989) or Box-Cox transformations are also possible.
[6] Strictly speaking there is also a 'scale' parameter, which reflects the variance of the random component of utility and is used for normalising the model. It is called 'scale' parameter, because it scales the β parameters in Eqs. (6.3a)−(6.3c); a higher random variance leads to lower estimated βs.

shipments (which may vary in terms of origins and destinations, leading to variation in distance and time within modes but over shipments). β_0 and β_4 are so-called 'alternative-specific constants', ASCs. There can only be N-1 ASCs in a model, N being the number of available choice alternatives, because in a utility maximisation model only differences in observed utility matter. In the example above, we have excluded an ASC for the inland waterways alternative, which means that for this alternative, the constant is normalised to 0. For the same reason, we can only include attributes as explanatory variables that differ between alternatives. Attributes of the decision-maker (e.g. the size of the firm) or of the shipment (e.g. containerised or not) can only be included by interacting these variables with characteristics of the modes (for instance by making certain firms less cost sensitive or including containerisation only for rail, expressing that container transports are more likely to be transported by rail).

Coefficients can be generic, such as β_1 for cost above, or alternative-specific, such as the other coefficients in Eqs. (6.3a)–(6.3c). Which is best is largely an empirical matter, which means that various forms should be tested and compared against each other. In the above model specification we have used generic coefficients for costs (but not for the other variables), which has the additional advantage that one unit of money paid for road transport has the same value as one unit of money paid for rail or inland waterways transport ('a euro is a euro' or 'a dollar is a dollar').

The decision on mode choice by decision-maker k could be influenced by other variables than the three attributes included above. For instance the flexibility of a mode, the service frequency and the probability of damage to the goods might also play a role. But the researcher that is constructing the mode choice model may not have any data on these influencing variables, or only data measured with some error. This is a key reason[7] for including the error components e_{road}, e_{rail} and e_{IWW} to the utility functions: they represent variables that affect the utility of the decision-maker, but are not observed by the researcher (or observed only with measurement errors).

In order to deal with the error components, the researcher assumes these are random variables (with a mean of zero and some variance). By making different specific assumptions on the probability distribution of the error components, different discrete choice models can be derived. These models are probability models, because they do not generate a certain choice, but probabilities for each of the available alternatives.

6.2.2   Different Distributional Assumptions Lead to Different Discrete Choice Models

The most common assumption for the error components e, both in passenger and freight modelling, is that they are independently and identically (i.e. same variance

---

[7] There are other reasons in the discrete choice literature for including the error terms.

across observations) distributed following the extreme value distribution type I (or Gumbel distribution). This leads to the MNL model with the choice probabilities:

$$P_{ik} = \frac{e^{G_{ik}}}{\sum_i e^{G_{ik}}}$$
(6.4)

The MNL model can be estimated by maximum likelihood methods that do not involve any simulation. Several software packages contain MNL estimation (sometimes called 'conditional logit').

After having estimated the model, one can apply the estimated coefficients on a sample of firms (shipments) to calculate probabilities for each choice alternative for each observation. If this sample is representative of the population studied, one can then simply sum the probabilities (this method is called 'sample enumeration') overall observations in the sample to get the market shares for the alternatives (e.g. the share of road transport in the total for toad, rail and inland waterways) as predicted by the model. For non-representative samples,[8] one can do a weighted summation with the population to sample fractions for each observation as weights. Different zones in a study area or different horizon years might even have different sets of weights. Such applications can give the impact of changing a single variable at a time (which can be expressed in the form of elasticities), but can also predict what would happen in case of an input scenario with changes for several (possibly all) variables in the model.

Discrete choice models estimated on stated preference (SP) data only should not directly be used for forecasting (including the derivation of elasticity values), since the SP data will have a different variance for the error component than real-world data, which will affect the choice probabilities. This happens because in the experimental set-up of the SP many things that can vary in reality are kept fixed, and vice versa. Therefore, for forecasting it is better to use revealed preference (RP) data or combined SP/RP data where the variance of the SP is scaled to that of the RP. Models estimated on SP data can directly be used to derive ratios of coefficients (such as a value of time or a value of reliability) because in calculating these ratios, the SP error component drops out.

A well-known restriction of the MNL model is that the cross-elasticities are the same: if in the mode choice model in Eqs. (6.3a)−(6.3c) the cost of road transport increases, substitution will occur to rail and inland waterways in proportion to their current market shares, so that the road cost elasticities of demand for rail transport and for inland waterway transport will be the same. Another manifestation of basically the same phenomenon (which is due to the independence of the error terms) is the independence from irrelevant alternatives (IIA) property: the ratio of the choice

---

[8] MNL models can be estimated consistently on a sample that is non-representative with regards to the exogenous variables. If the sample is non-representative with regards to the choice variable (e.g. with an overrepresentation of rail transport), and the model has N-1 ASCs and M other coefficients, all M coefficients can still be estimated consistently using standard methods, and only the N-1 ASCs will be biased. These ASCs can simply be corrected after estimation on the basis of the observed market shares (McFadden, 1981).

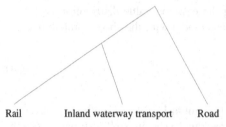

Figure 6.1 Nested logit structure for freight mode choice.

Rail   Inland waterway transport   Road

probabilities between two alternatives does not depend on any other alternative. These properties may be at odds with reality. In practice, there could for instance be more substitution between rail and inland waterways than between any of these alternatives and road transport. A relatively easy way to accommodate for this is the nested logit model (e.g. Daly & Zachary, 1978) in which rail and inland waterways would be grouped in a nest, allowing correlation between these alternatives.

Mathematically, the easiest representation is to distinguish two probabilities (as for instance in Train, 2003), linked to each other by the logsum variable (Figure 6.1).

$$P_{B_l k} = \frac{e^{G_{lk} + \lambda_l I_{lk}}}{\sum_l e^{G_{lk} + \lambda_l I_{lk}}} \tag{6.5a}$$

$$P_{\langle ik | B_l \rangle} = \frac{e^{H_{ik}/\lambda_l}}{\sum_{i \in B_l} e^{H_{ik}/\lambda_l}} \tag{6.5b}$$

$$I_{lk} = \ln \sum_{i \in B_l} e^{H_{ik}/\lambda_l} \tag{6.5c}$$

The first probability Eq. (6.5a) gives the chance that decision-maker k chooses an alternative within nest B_l. This depends on the generalised cost G of l plus a coefficient λ_l times the expected cost from the alternatives in the nest, represented by I_{lk}, the so-called 'logsum' variable, relative to the same kind of costs for the all alternatives at the nesting level.

The second (conditional) probability gives the change of choosing alternative i given that nest B_l has been chosen. This depends on the generalised cost of this alternative relative to those for all alternatives in the nest.

Now the unconditional probability that decision-maker k will select alternative i is:

$$P_{ik} = P_{\langle ik | B_l \rangle} P_{B_l k} \tag{6.5d}$$

The coefficient λ_l is the 'logsum coefficient' which gives the degree of correlation between the error components of the alternatives in nest B_l: the higher this coefficient, the lower the correlation. In estimation, this is an extra parameter to be estimated. The estimated value must be between 0 and 1 for global consistency (meaning: across the entire range for the exogenous variables) with RUM. If a

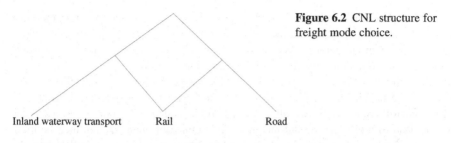

Figure 6.2 CNL structure for freight mode choice.

Inland waterway transport   Rail   Road

value above 1 is found, this often is an indication that a different (especially a reversed) nesting structure would work better and be consistent with RUM.

Both MNL and nested logit are members of a family of models, the GEV (Generalised Extreme Value) family (Daly & Bierlaire, 2006; McFadden, 1978) which contain more members (and sometimes new members are discovered), all of which are consistent with RUM. Most of these have only seen a limited number of applications in passenger transport and none or almost none in freight transport, though they offer more flexibility in terms of substitution patterns between alternatives than MNL or nested logit. This includes cross-nested logit (CNL) (Ben-Akiva & Bierlaire, 1999; Wen & Koppelman, 2001) where a single alternative can belong to more than one nest at the same time. A potential example is given in Figure 6.2 below, where rail is in a nest with inland waterway transport (as before), and also in another nest with the other non-water-based mode, road transport. Inland waterway transport and road transport are not correlated with each other.

The corresponding equations for CNL are:

$$P_{B_l k} = \frac{e^{G_{lk} + \lambda_l I_{lk}}}{\sum_l e^{G_{lk} + \lambda_l I_{lk}}} \tag{6.6a}$$

$$P_{\langle ik | B_l \rangle} = \frac{\alpha_{il} e^{H_{ik}/\lambda_l}}{\sum_{i \in B_l} \alpha_{il} e^{H_{ik}/\lambda_l}} \tag{6.6b}$$

$$I_{lk} = \ln \sum_{i \in B_l} \alpha_{il} e^{H_{ik}/\lambda_l} \tag{6.6c}$$

$$P_{ik} = \sum_l P_{\langle ik | B_l \rangle} \cdot P_{B_l k} \tag{6.6d}$$

In Eqs. (6.6a)−(6.6d), l now stands for a specific nesting structure for alternative i (for instance 'slow modes nesting' and 'land modes nesting'). There now is an extra parameter α_{il}: the share of the lth nesting structure for alternative i (the degree of membership of the nest). Overall nesting structures l, the α_{il} should sum to 1. When one would have two nesting structures (as in the example above), the standard procedure is to give both a weight of ½. Hess, Fowler, Adler, & Bahreinian (2009) also used ½ and remarked that estimation of these membership coefficients would have been very difficult.

Yet another GEV (Generalised Extreme Value) model, the ordered extreme value OGEV (Ordered Generalised Extreme Value) model (Small, 1987) is not interesting for mode choice, because it presupposes ordered alternatives (e.g. time periods during one day), whereas mode choice is unordered.

Assuming that the error components follow a multivariate normal distribution leads to the probit model. This model is almost as straightforward to estimate as MNL when there are only two alternatives. Probit models with more than two alternatives lead to multidimensional integrals, and simulation methods can then be used for estimation (e.g. Geweke, Keane, & Runkle, 1994), but this is all (still) relatively cumbersome. On the other hand, the probit model is very flexible, since one can define and estimate specific correlation parameters between the alternatives. Winston (1981) used a probit model for the binary choice between road and rail transport in the United States. Tavasszy (1996) also used the probit model for an application to mode choice for the transport corridor from the UK to the Netherlands.

The mixed logit model was developed in the late nineties and saw its breakthrough in the first decade of the twenty-first century. It has two error components $e_{ik} + v_{ik}$, where one follows the Gumbel distribution (so that conditional on the other, the model is MNL) and the other can follow any distribution, such as the normal or lognormal. Mixed logit can accommodate two ideas, which are both generalisations of MNL:

1. Random coefficients or taste variation: the coefficients, such as β_1 above, follow a statistical distribution of which the mean and the variance are estimated. This is potentially very important for freight mode choice models because we might expect that the heterogeneity in freight transport carries over to coefficients like the transport time and cost coefficients in mode choice models. It is even possible to include interaction coefficients for observed heterogeneity and randomness for unobserved heterogeneity in coefficients like those for transport time and cost at the same time. It is also possible in mixed logit to estimate several random coefficients (e.g. for transport cost and time) where in taking the random draws one takes account of the fact that draws for the same individual should not be independent. This is important for panel data or for stated preference experiments where the respondents are asked to make several choices. With such data sets, the researcher cannot assume that the observations are independent.

2. Error components: specific correlation structure between alternatives. In its most general form, the utility function now becomes:

$$U_{ik} = \sum_r \beta_r x_{rik} + \sum_s \sum_t \eta_s w_{st}^i v_t + e_{ik} \tag{6.7}$$

In which:

The first term on the right-hand side denotes the influence of the usual observed attributes on utility.

The second term on the right-hand side determines the error component structure:

- v_t is an error component, following some statistical distribution $f(0,1)$, which can consist of several random subcomponents ($t = 1,\ldots,T$).
- η_s is a coefficient to be estimated.
- w_{st}^i is a general weighting matrix, based on data and/or fixed by the researcher, for alternative i, with rows s corresponding to the coefficients η and columns t corresponding to the error subcomponents in v.

The third term on the right-hand side e_{ik} is the error component that in the mixed logit model is Gumbel distributed (if v_t and e_{ik} would follow the multivariate normal distribution, the model would be multinomial probit).

Estimation of the model requires simulation (Train, 2003) in the sense of taking a large number of random draws for v and calculating the likelihood function for each of the draws, but because of the presence of the Gumbel component this can be done much faster than for probit.

Many models in the scientific journals in transport, especially in passenger transport, nowadays use mixed logit. Often the estimation data are stated preference data and the purpose of the estimation is to provide monetary values for non-monetary attributes, such as transport time (value of travel time), reliability (value of travel time variability), as will be discussed in Chapter 10. The SP alternatives in these experiments and models are usually 'abstract' or 'unlabelled' alternatives, meaning that it is not specified whether the choice alternatives are different modes, route, carriers, etc. Such studies only require that the mixed logit simulation of the non-Gumbel distribution (e.g. drawing many times from a lognormal distribution) is done for each model specification that is estimated. For transport models that are developed for use in forecasting for scenarios and for the impact of transport projects and policies relative to a reference scenario, mixed logit is still not very attractive because every time the model is applied, the random draws need to be made again, which is very time consuming. In the framework of a large model system (e.g. with many zone pairs), possibly with feedback loops, this would simply take too long. Hardly any practical transport forecasting model, in passenger and freight transport, therefore uses mixed logit.[9]

An important limitation of mixed logit with continuous statistical distributions for both error components is that the researcher has to assume the shape of the distributions (e.g. normal or lognormal distribution) and this assumption could strongly affect the final result. In addition, this technique does not always yield a stable estimation result. A better choice could then the latent class model that is related to mixed logit. Within the random coefficients specification, this model uses a discrete number of possible values (latent classes) for a coefficient like the cost coefficient. Furthermore, membership equations can be estimated, linking membership to one of the classes to observed variables (which can include specially collected attitudinal variables, but this is not necessary). In Hess, Ben-Akiva, Gopinath, and Walker (2011), latent class models are compared against mixed logit models with a continuous distribution, with a positive outcome for latent class models. Similar conclusions on latent class models were reached in Greene & Hensher (2012).

6.2.3   Non-RUM Models

By interpreting discrete choice models as RUM models, these models are founded in microeconomic theory and represent 'rational' behaviour. However, the economic perspective is just one way of looking at the behaviour of humans or firms,

[9] An exception, albeit in passenger transport, is the SILVESTER model for Stockholm.

and there are other aspects of behaviour (e.g. as studied by psychology or sociology) as well. For some decisions (the economic perspective might be more relevant (here one might think of important well-planned long-run decisions, especially in a business context), for other decisions other perspectives might be more important. For a long time, RUM was the only available serious candidate as a foundation for mathematical models of making discrete choices, but in recent years two other paradigms have gained some popularity in transport modelling: prospect theory and regret minimisation.

Prospect theory (Kahneman & Tversky, 1979, 1992; van der Kaa, 2008) was developed by psychologists, who found that in many experimental situations, respondents did not follow the theory of rational behaviour, and who then tried to stay closer to how people actually respond in these experiments. Prospect theory says that:

- The valuation of an attribute depends on the present value of that attribute, i.e. it depends on the reference alternative (the situation as observed now): reference dependence.
- There will be a difference in the valuation of gains and losses in an attribute: losses loom larger, per unit: loss aversion.
- There will be a difference in the valuation at different values of an attribute (e.g. between a short and a long transport): size dependence.
- There is non-linear weighting of probabilities by respondents (Hensher & Li, 2012; Koster & Verhoef, 2012). An application to SP data in freight transport (but for abstract alternatives, not modal choice) can be found in Halse, Samstad, Killi, Flügel, & Ramjerdi (2010).

Instead of utility maximisation, discrete choice models can also be based on the minimisation of anticipated regret by the decision-makers. Regret occurs when a non-chosen alternative scores better on some attribute than the chosen alternative. When two regret components are distinguished in the regret function, 'observed' or 'systematic' regret and random regret, this leads to the random regret minimisation (RRM) model (Chorus, 2010). Systematic regret is calculated by doing a pairwise comparison of available alternatives on the basis of the attribute values of these two alternatives, and then summing the outcomes overall pairs.

The distributional assumptions for the random component(s) can be similar to those of the RUM model, and even the same software can be used for RRM and RUM, but the choice probabilities from the two types of models will be different. The key difference here is that RRM captures the 'compromise effect': alternatives that perform 'in-between' on all attributes, relative to the other alternatives, are generally favoured over alternatives with a poor performance on some attributes and a strong performance on others (Chorus & de Jong, 2011).

6.2.4 Interaction Between Agents in Freight Transport

One of the defining characteristics of freight transport is that several decisions-makers are involved for the same shipment (though for a particular shipment, one decision-maker may dominate). For the choice of mode, shippers (as senders) can be important, but the requirements of the receiver will also play a role. In a number of supply chains (e.g. food products) the receivers (in this example the large supermarket retailers) are the ones organising the transport. There are also situations where shippers and receivers leave it to the carriers to select the mode. Several logistics service providers nowadays offer a range of different door-to-door transport solutions and are happy to make the choice among these for their clients (for a fee).

Nevertheless, this commonly acknowledged observation that mode choice (and other choices in freight transport) might be determined in interactions between several agents (senders, receivers and carriers) has not (yet) led to many studies that analyse decision-making in freight in the form of interactions between various parties. This is probably due to the complications that arise when one distinguishes several interacting decision-makers for the same decision. Nevertheless some progress has been made in the past decade on modelling interactions between decision-makers in freight transport.

One line of approach to modelling interactions is the combination of game theory with experimental economics to collect data. Such economic experiments are often carried out with students in the role of economic agents that interact with each other in various market market settings. Holguín-Veras, Thorson, & Ozbay (2004) studied the formation of truck tours with various carriers. The already-mentioned Holguín-Veras et al. (2011) study used experimental economics (with students as participants) to study interactions between shippers and carriers in mode and shipment size choice, coming to the conclusion that in a competitive transport service market, the shippers and the carrier will cooperate in the selection of mode and shipment size, irrespective of who is leading, the shippers or the carriers. An overview of experimental economics applications and ideas on its future potential can be found in de Jong (2012).

Another line of research tries to integrate the idea of interactions into stated preference survey design and modelling. Hensher (2002) introduced such stated preference (SP) experiments and described how discrete choice models can be extended to include interactions. Hensher first developed interactive agency choice experiments (IACE). These SP experiments include sequential choices, where agents are informed about the previous choices of other agents. This is done for pairs (or 'dyads') of agents (e.g. shipper-carrier) and the process of feeding back information continues for all pairs where agreements have not been reached. Alternatives and choice sets can be correlated within and between agents.

IACE have been carried out in passenger transport, but are particularly applicable in freight transport (as argued in the beginning of this Section 6.2.4), and the interaction between shippers and carriers has indeed been analysed using IACE and econometric models for estimation on these choice data of the type discussed in Section 6.2 (Hensher & Puckett, 2005).

A serious disadvantage of IACE is that the survey costs of interviewing shippers on the responses of the carriers etc. can be quite high (compared to more standard stated preference surveys).The resulting samples will therefore be small. This led to the development of minimum information group inference, MIGI

(Puckett & Hensher, 2006), where agents are interviewed only once. The innovation here is that in the choice modelling each shipper is matched with a carrier, and their group decision-making is inferred.

6.3 Practical Examples

6.3.1 Aggregate Mode Choice Models

As discussed in Section 6.2, the standard disaggregate choice model (MNL), and many extensions of it, can be based on the theory of utility maximisation by individual decision-makers. There also is an aggregate form of this model (often called 'aggregate logit') where the observations usually refer to summations of shipments for the same OD zone pair.[10] More specifically these modal split models are estimated on data for the market share of each mode over different OD pairs. The aggregate modal split model can indirectly be based on the theory of individual utility maximisation (all decision-makers on the shipments for an OD pair carrying optimising their subjective utility, but only under very restrictive assumptions. These assumptions basically boil down to assuming that all variation in characteristics of the decision-makers and of the goods belongs to the error component of the utility function. This would be such a far-reaching assumption, that it is better to see aggregate logit models as pragmatic models (that have shown to be able to yield plausible results) instead of models based on a theory of behaviour.

The aggregate logit models are often selected because disaggregate data are not available and all we have are the tonnes by OD zone pair and mode (or tonnes by PC pair and main mode). The aggregate logit can easily be estimated (both with software for linear regression models and for discrete choice models), produces an intuitively appealing S-shaped market shares curve and market shares, which are always between 0 and 1. Because of these advantages, the aggregate logit model still is the single model specification used most in practical freight mode choice modelling.[11]

A typical formulation is the 'difference form'[12]:

$$\log \frac{S_i}{S_j} = \beta_0 + \beta_1(P_i - P_j) + \sum_w \beta_w(x_{iw} - x_{jw}) \tag{6.8}$$

---

[10] The observations might also be summations of shipments to form data per business sector or time series for some region.

[11] In practice it is often even difficult to obtain plausible transport time and costs coefficients when estimating on aggregate data. Prof. Moshe Ben-Akiva once suggested here to assume a value of time distribution to allow for heterogeneity between shipments.

[12] An alternative for the difference form is the 'ratio form' where the right-hand side has P_i/P_j and x_{iw}/x_{jw}, which has the disadvantage that the choice of the base mode (in the denominator of the dependent variable) affects the estimation results and the elasticities from the model. The difference form does not have this disadvantage.

In which:

S_i/S_j is the ratio of the market share of mode i to the market share of mode j.
P_i and P_j are the transport costs using these two modes.
$X_{iw} - X_{jw}$ are w ($w = 1,\ldots, W$) differences in other characteristics of the two modes.

This model can be estimated using specialised discrete choice estimation software (the same as used for estimating disaggregate models), but also by standard regression analysis of the log-ratio above on its explanatory variables.

Aggregate modal split models are mostly binomial (two available modes) or MNL models (three or more available modes). Since they only give the market share of a mode, not the absolute amount of transport (tonnes) or traffic (vehicles), the elasticities from such models are conditional elasticities (conditional on the quantity demanded; see Beuthe, Geerts, & 'Koul à Ndjang' Ha', 2001).

Examples of aggregate model split models are:

- Blauwens & van de Voorde (1988) modelled the choice of inland waterways versus road transport in Belgium.
- The modal split model within the NEAC model for Europe (NEA, 2000) explains the choice between road, rail and inland waterway transport.
- The French national freight transport model MODEV (MVA and Kessel + Partner, 2006) has an aggregate logit model with road, rail, combined road–rail and inland waterways as modal alternatives.
- The current EU transport model system Transtools (Tetraplan, 2009) includes a modal split component for freight by road, rail and inland waterway transport that is an aggregate logit model.
- The LEFT model (Fowkes, Johnson, & Whiteing, 2010) for the choice between road and rail for seven commodities and nine distance bands also works at the aggregate level.
- The new 'back to basics' Dutch freight transport model BasGoed also is an aggregate logit model, estimated on shares by zone pair for the modes road, rail and inland waterway transport (de Jong et al., 2011). The level-of-service inputs come from uni-modal assignments for these modes.
- The following aggregate models in a way go beyond the aggregate logit model in that they model the budget share of a mode in total transport cost. This type of input demand function can be derived from a production cost function for a firm that also includes the cost of transport services by mode, using Shephard's Lemma from the standard microeconomic theory of the firm. This too however, is a relation that applies for a single firm; transfer to a sector or region is not straightforward.
  - Oum (1979) had aggregate time series data on modal split in Canada and estimated various aggregate models on this.
  - Friedlaender & Spady (1980) analysed the mode choice at the level of 96 economic sectors in five regions in the United States (so the data are not by OD pair, but by sector and origin region).
  - Oum (1989) used aggregate models with different cost specifications (linear, loglinear, Box-Cox and translog) explaining the modal split for transport flows between Canadian regions.

6.3.2 Disaggregate Mode Choice Models

A practical example of an *MNL model* estimated on disaggregate freight data for mode choice is Nuzzolo & Russo (1995). This is the mode choice model within the Italian national freight model system for intercity freight flows. It contains three choice alternatives (road, rail and combined road—rail transport) and was estimated on interview data (RP) with producers/shippers. This model includes transport costs and time by mode as well as some shipment characteristics.

Nested logit can also be used to combine two choices in a joint model (such as shipment size and mode choice, or mode and supplier choice) and for joint estimation on a combination of data sets (e.g. stated preference and revealed preference data, Bradley & Daly, 1997). Two practical examples of nested logit mode choice models in freight transport are:

- Jiang et al. (1999): a model for the choice of mode (more specifically own account transport versus a nest with three contract out options: road, rail and combined road—rail transport) estimated on the French shippers survey of 1988. The model includes attributes of the firms and of the shipment, but not transport time and cost (distance however was included).
- de Jong, Vellay, & Houée (2001) used RP information and data from SP mode choice experiments (and from SP abstract or 'within-mode' alternatives) among shippers in the French region Nord-Pas de Calais to estimate a mode choice model. Nested logit was used to allow simultaneous SP/RP estimation. The explanatory variables in the model include transport time and costs by model, reliability, flexibility and frequency of the mode, as well as attributes of the shipments and the shipper.
- The German model for federal infrastructure project assessment (BVWP model) was estimated as a disaggregate mode choice model (road, rail and inland waterway transport), partly on stated preference data (ITP & BVU, 2007).

A rare example of a CNL model in freight transport mode choice is Arunotayanun (2009), who estimated CNL models on two different databases: (1) on an SP survey among shippers in Indonesia, where the choice alternatives are small truck, large truck and rail; (2) on the French RP data set ECHO on individual shipments (also see Chapter 5), using road-own account, road-contract out, rail and combined road—rail as choice alternatives and testing for different cross-nested structures.

Examples of application of the *mixed logit model* to freight mode choice are:

- Using data from an SP survey among shippers in Italy, Massiani (2008) estimated a random coefficients model for the choice between road transport and combined road—rail transport, with random coefficients for transport time, cost, reliability (measured as the percentage on time), flexibility, probability of damage and the model constant. He tested a normal distribution for these random coefficients as well as a triangular, and also included interaction variables for attributes of the sector and the shipment.
- Ben-Akiva, Bolduc, & Park (2008) used data from an SP/RP survey of 166 shippers in the United States to estimate random coefficients models for the choice between truck, rail and combined road—rail transport. They included not just transport costs but also capital cost, storage cost, safety stock cost and cost of loss and damage, effectively

minimising total logistics cost. They also include a number of latent attitudinal variables to explain modal flexibility, which in turn is one of the explanatory variables in mode choice.

- Arunotayanun (2009) estimated mixed logit models on the Indonesian SP data set mentioned above. Random coefficients are used for transport time, transport cost, reliability and other SP attributes.
- Fries, de Jong, Patterson, & Weidmann (2010) carried out SP surveys among shippers in Switzerland and estimated MNL and mixed logit models on the resulting SP data. Three modes are distinguished: road, rail and combined road−rail. Four of the attributes from the SP experiments (transport time and costs, reliability measured as percentage delivered on time and the greenhouse gas emissions) are treated as random coefficients, estimating a mean coefficient as well as a standard deviation for three assumptions on the distribution of the random coefficients: Normal, lognormal and triangular.
- The choice between road and rail transport in Spain for the inland leg of containerised maritime shipments is modelled by Feo-Valero, Garcia-Menendez, Saez-Carramolino, & Furio-Prunonosa (2011). The database is a stated preference survey among shippers and the model used is mixed logit with a single individual-specific component to take account of the fact that the data contain several observations for the same respondent.

Practical applications of *latent class modelling* to the choice of mode in freight transport are:

- Gopinath (1995) used two classes of coefficient values to capture some of the heterogeneity in tastes of shippers in a model estimated on combined SP/RP data (and additional attitudinal questions) in the United States.
- Arunotayanun (2009) also estimated latent class models on the Indonesian SP data mentioned above.

Most applications of *RRM* so far have been in passenger transport (or in health or environmental economics), but there also is a recent application of RRM to freight transport mode choice (Boeri & Masiero, 2012), where RRM outperformed RUM.

6.3.3   Joint Models for Mode Choice and Related Choices

Some practical disaggregate freight transport models that simultaneously deal with *mode choice and logistic choices* are the following.

1. Models with discrete mode (or transport chain) choice, jointly with discrete shipment size choice (after dividing shipment size into a number of classes):
   - Chiang, Roberts, & Ben-Akiva (1981) modelled the choice of shipment size, mode and the location of supplier (supply zone).
   - de Jong (2007) and de Jong & Ben-Akiva (2007), using data from the Swedish commodity flow survey (CFS) 2001, estimated a joint model of mode and shipment size choice.
   - Habibi (2010) estimated models for discrete shipment sizes and transport chains for the commodity 'domestic steel products' on the CFS 2004/2005.
   - Windisch, de Jong, & van Nes (2010) used the CFS 2004/2005 to estimate models of (discrete) shipment size and transport chain choice.

These models have as choice alternatives combinations of a mode m ($i = 1, \ldots$, I) and a shipment size class s ($s = 1, \ldots$, S), with a choice set that could be as large as I.S alternatives. The econometric models used are MNL, nested logit (with more substitution between shipment size classes than between modes) and mixed logit.

2. Models with discrete mode choice jointly with continuous shipment size choice:
   - McFadden, Winston, & Boersch-Supan (1985) developed a model for shipment size and mode choice and applied it to agricultural goods.
   - Abdelwahab & Sargious (1992) and Adelwahab (1998) estimated a mode choice and shipment size model on the US Commodity Transportation Survey.
   - de Jong & Johnson (2009) and Johnson & de Jong (2011) used the CFS 2001 to estimate discrete—continuous (continuous shipment size) models, following the specification that Holguín-Veras (2002) developed for road vehicle type and shipment size choice. In these papers, discrete choice models (both mode and shipment size treated as discrete variables) were estimate on the same data, allowing a comparison of both models.
   - Combes (2010a, 2010b) developed models of shipment size and mode choice on the national French shippers survey ECHO.
   - Liu (2012) estimated models for discrete mode and continuous shipment size in the CFS 2001 for four different commodity groups.

   These discrete—continuous models are usually estimated in two steps (one for the discrete and one for the continuous step) with selectivity correction terms to correct for the simultaneity bias.

Models that include the *joint choice of mode and supplier* are:

- The model by Chiang et al. (1981) also falls in this category, since the dependent variables are shipment size, mode and the location of supplier (represented by supply zone).
- Another disaggregate model where a receiver of the goods chooses the supplier to buy from is Samimi, Mohammadian, & Kawamura (2010). Their utility function includes receiver characteristics (quantity required, budget and modes) and supplier characteristics (capacity to produce/stock, price and geographic location).

Examples of *disaggregate joint mode and route* choice (both discrete variables) models are:

- The freight transport model developed for the Öresund screenline between Sweden and Denmark (e.g. Fosgerau, 1996), where three types of trucks, unaccompanied trailer, rail, and intermodal road—rail were combined with different ferry routes and a fixed link route, using a combination of SP and RP data.
- A similar model for the Fehmarn Belt corridor between Denmark and Germany (Fehmarn Belt Traffic Consortium, 1998), also on SP and RP data.

The *aggregate mode and route choice model (multi-modal assignment)* has been used in a number of cases:

- Even in a relatively small network, many route-mode combinations can be chosen for a specific OD combination and a cost minimisation algorithm is used to find the least-cost combination. The cost function that is minimised in multi-modal assignment can contain several attributes, including transport time components and terminal cost. In most cases all traffic for an OD pair is assigned to the single optimal alternative: all-or-nothing assignment, but for instance the Dutch SMILE + model (Tavasszy, Smeenk, & Ruijgrok (1998)) uses stochastic multi-modal assignment.

- One of the commercial software packages for multi-modal network assignment is the STAN package (Crainic, Florian, Guelat, & Spiess, 1990), which has been used in the previous freight transport models in Norway (NEMO) and Sweden (the previous SAMGODS model), and also in Canada and Finland. The WFTM freight model for the Walloon Region uses a similar multi-modal network assignment, but this is implemented in the NODUS software (Beuthe et al., 2001; Geerts & Jourguin, 2000). The selection of the optimal mode-route combination is done separately for different commodity groups, because different goods will have different handling requirements and values of time, and therefore the coefficients in the cost functions (e.g. for transshipment costs and time costs) will differ between these goods.

In the European model SCENES (SCENES Consortium, 2001) and the Great Britain Freight Model (GBFM) a multi-modal network assignment takes place. Worldnet (Newton, 2008), that was also developed for the European Commission, and covers Europe, but also intercontinental sea and air freight, contains a multi-modal transport chain builder. Furthermore, it also includes an aggregate logit model to choose between the different uni-modal and multi-modal transport chains.

References

Abdelwahab, W. M. (1998). Elasticities of mode choice probabilities and market elasticities of demand: evidence from a simultaneous mode choice/shipment-size freight transport model. *Transportation Research E,*(34), 257−266.

Abdelwahab, W. M., & Sargious, M. A. (1992). Modelling the demand for freight transport. *Journal of Transport Economics and Policy, 26*(1), 49−70.

Arunotayanun, K. (2009) *Modelling freight supplier behaviour and response* (Ph.D. thesis). London: Centre for Transport Studies, Imperial College.

Ben-Akiva, M., & Bierlaire, M. (1999). Discrete choice methods and their applications to short-term travel decisions. In R. Hall (Ed.), *Handbook of transportation science*. Amsterdam: Kluwer.

Ben-Akiva, M., Bolduc, D., & Park, J. Q. (2008). Discrete choice analysis of shipper's preferences. In M. Ben-Akiva, H. Meersman, & E. van de Voorde (Eds.), *Recent developments in transport modelling: Lessons for the freight sector*. Bingley, UK: Emerald.

Beuthe, M. B. J., Geerts, J. -F., & Koul a Ndjang Ha, C. (2001). Freight transport demand elasticities: a geographic multimodal transportation network analysis. *Transportation Reseach E, 37*, 253−266.

Blauwens, G., & Van de Voorde, E. (1988). The valuation of time savings in commodity transport. *International Journal of Transport Economics, XV*(1), 77−87.

Boeri, M., & Masiero L. (2012). Regret minimization and utility maximization in a freight transport context: an application from two stated choice experiments. In: *Swiss transport research conference,* Monte Verità, 2−4 May 2012.

Bradley, M. A., & Daly, A. J. (1997). Estimation of logit choice models using mixed stated preference and revealed preference information. In P. R. Stopher, & M. Lee-Gosselin (Eds.), *Understanding travel behaviour in an era of change*. Amsterdam: Elsevier.

Chiang, Y., Roberts, P. O., & Ben-Akiva M. E. (1981). Development of a policy sensitive model for forecasting freight demand, Final report, Center for Transportation Studies Report 81-1, MIT, Cambridge, MA.

Chorus, C. G. (2010). A new model of random regret minimization. *European Journal of Transport and Infrastructure Research*, *10*(2), 181−196.

Chorus, C. G., & de Jong, G. C. (2011). Modeling experienced accessibility: an approximation. *Journal of Transport Geography*, *19*, 1155−1162.

Combes, P. F. (2010a). The choice of shipment size in freight transport PhD, Paris: Université Paris-Est.

Combes, P. F. (2010b). Estimation of the economic order quantity model using the ECHO shipment data base. In: *European transport conference*, Glasgow.

Crainic, T. G., Florian, M., Guelat, J., & Spiess, H. (1990). Strategic planning of freight transportation: STAN, an interactive-graphical system. *Transportation Research Record*, *1283*, 97−124.

Daly, A., & Bierlaire, M. (2006). A general and operational representation of GEV models. *Transportation Research B*, *40*, 285−305.

Daly, A. J., & Zachary, S. (1978). Improved multiple choice models. In D. A. Hensher, & M. Q. Dalvi (Eds.), Determinants of travel choice. Saxon House, Sussex.

de Jong, G. C. (2007). A model of mode and shipment size choice on the Swedish commodity flow survey. In: *UTSG 2007*, Harrogate.

de Jong, G. C.(2012). Application of experimental economics. In: *Transport and logistics*, European Transport/Trasporti Europei, Issue 50, Paper 3.

de Jong, G. C., & Ben-Akiva, M. (2007). A micro-simulation model of shipment size. *Transportation Research B*, *41*(9), 950−965.

de Jong, G. C., Burgess, A., Tavasszy, L.,Versteegh, R., de Bok. M., & Schmorak, N. (2011). Distribution and modal split models for freight transport in The Netherlands. In: *European transport conference 2011*, Glasgow.

de Jong, G. C., & Johnson, D. (2009). Discrete mode and discrete or continuous shipment size choice in freight transport in Sweden. In: *European transport conference*, Noordwijkerhout, The Netherlands.

de Jong, G. C., Vellay. C., & Houée, M. (2001). A joint SP/RP model of freight shipments from the region Nord-Pas de Calais. In: *2001 European transport conference*, Cambridge.

Fehmarn Belt Traffic Consortium. (1998). Fehmarn Belt traffic demand study, Final Report; FTC (Carl Bro, ISL, BVU, HCG en ITP), Copenhagen-Bremen-Freiburg-Den Haag-München.

Feo-Valero, M., Garcia-Menendez, L., Saez-Carramolino, L., & Furio-Prunonosa, S. (2011). The importance of the inland leg of containerised maritime shipments: an analysis of modal choice determinants in Spain. *Transportation Research E*, *47*(4), 446−460.

Fosgerau, M. (1996). *Freight traffic on the Storebælt fixed link. Proceedings of seminar D&E*, 24th European transport forum. London: PTRC.

Fowkes, A. S., Johnson, D. & Whiteing, A. E. (2010). Modelling the effect of policies to reduce the environmental impact of freight transport in Great Britain. In: *European transport conference*, Glasgow.

Friedlaender, A. F., & Spady, R. (1980). A derived demand function for freight transportation. *Review of Economics and Statistics*, *62*, 432−441.

Fries, N., de Jong, G. C., Patterson, Z., & Weidmann, U. (2010). Shipper willingness to pay to increase environmental performance in freight transportation. *Transportation Research Record* (2168), 33−42.

Geerts, J. F., & Jourguin G. (2000). Freight transportation planning on the European multimodal network: the case of the Walloon region. In: *Proceedings of the 6th RSAI world congress,* Lugano, Switzerland.

Geweke, J., Keane, M., & Runkle, D. (1994). Alternative computational approaches to inference in the multinomial probit model. *Review of Economics and Statistics, 76,* 609–632.

Gopinath, D. A. (1995) *Modeling heterogeneity in discrete choice processes: application to travel demand* (Ph.D. thesis). Cambridge, MA: MIT.

Greene, W. H., & Hensher, D. A. (2012). Revealing additional dimensions of preference heterogeneity in a latent class mixed multinomial logit model. *Applied Economics, 2012,* 1–6.

Habibi, S. (2010). *A discrete choice model of transport chain and shipment size on swedish commodity flow survey.* Stockholm: Kungliga Tekniska högskolan, Department of Transport and Logistics.

Halse, A., Samstad, H., Killi, M., Flügel, S., & Ramjerdi, F. (2010). Valuation of freight transport time and reliability (in Norwegian), TØI report 1083/2010, Oslo.

Hensher, D. A. (2002). Models of organisational and agency choices for passenger and freight-related travel choices: notions of inter-activity and influence. In: *Resource paper prepared for the 8th IATBR conference workshop on 'models of organisational choices (freight and passenger)',* Lucerne.

Hensher, D. A., & Puckett, S. M. (2005). Refocussing the modelling of freight distribution: development of an economic-based framework to evaluate supply chain behaviour in response to congestion charging. *Transportation, 32,* 573–602.

Hensher, D. A., & Li, Z. (2012). Valuing travel time variability within a rank-dependent utility framework and an investigation of unobserved taste heterogeneity. *Journal of Transport Economics and Policy, 46*(2), 293–312.

Hess, S., Ben-Akiva, M., Gopinath, D. & Walker, J. (2011). Advantages of latent class over continuous mixture of logit models, Working paper, ITS, University of Leeds.

Hess, S., Fowler, M., Adler, T. & Bahreinian, A. (2009). A joint model for vehicle type and fuel type. In: *European transport conference,* Noordwijkerhout, The Netherlands.

Holguín-Veras, J. (2002). Revealed preference analysis of the commercial vehicle choice process. *Journal of Transportation Engineering, American Society of Civil Engineers, 128*(4), 336–346.

Holguín-Veras, J., Thorson, E., & Ozbay, K. (2004). Preliminary results of an experimental economics application to urban goods modeling research. *Transportation Research Record, 1873,* 9–16.

Holguín-Veras, J., Xu, N., de Jong, G. C., & Maurer, H. (2011). An experimental economics investigation of shipper-carrier interactions on the choice of mode and shipment size in freight transport. *Networks and Spatial Economics, 11,* 509–532.

ITP, BVU (2007). Prognose der deutschlandweiten Verkehrsverflechtungen 2025, ITP/BVU, München/Freiburg.

Jiang, F., Johnson, P., & Calzada, C. (1999). Freight demand characteristics and mode choice: an analysis of the results of modelling with disaggregate revealed preference data. *Journal of Transportation and Statistics, 2*(2), 149–158.

Johnson, D., & G. C., de Jong (2011) Shippers' response to transport cost and time and model specification in freight mode and shipment size choice. In: *2nd International choice modelling conference,* Leeds.

Kahneman, D., & Tversky, A. (1979). Prospect theory: an analysis of decision under risk. *Econometrica, 47,* 263–291.

Kahneman, D., & Tversky, A. (1992). Advances in prospect theory: cumulative representation of uncertainty. *Journal of Risk and Uncertainty*, 5(4), 297–323.

Koster, P., & Verhoef, E. T. (2012) A rank dependent scheduling model. *Journal of Transport Economics and Policy*, 46(1), 123–138.

Liu, X. (2012) Estimating value of time savings for freight transport: a simultaneous decision model of transport mode choice and shipment size, Örebro University Business School, Department of Economics.

Maibach, H., Schreyer, C., Sutter, D., van Essen, H. P., Boon, B. H., Smokers, R., et al. (2008) Handbook on estimation of external costs in the transport sector, report produced within the project Ínternalisation measures for all external costs of transport (IMPACT), CE Delft, Delft.

Massiani, J. (2008) Explaining, modelling and measuring the heterogeneity in shipper's value of time, Università degli studi di Trieste, Dipartimento di Scienza Economica e Statistica.

McFadden, D. (1974). Conditional logit analysis of qualitative choice-behaviour. In P. Zarembka (Ed.), *Frontiers in econometrics* (pp. 105–142). New York, NY: Academic Press.

McFadden, D. (1978). Modelling the choice of residential location. In A. Karlqvist, L. Lundqvist, F. Snickars, & J. Weibull (Eds.), *Spatial interaction theory and residential location*. Amsterdam: North-Holland.

McFadden, D. (1981). Econometric models of probabilistic choice. In C. Manski, & D. McFadden (Eds.), *Structural analysis of discrete data: With econometric applications*. Cambridge, MA: The MIT Press.

McFadden, D. L., Winston, C., & Boersch-Supan, A. (1985). Joint estimation of freight transportation decisions under non-random sampling. In E. F. Daughety (Ed.), *Analytical studies in transport economics*. Cambridge: Cambridge University Press.

MVA and Kessel + Partner (2006) MODEV–Modèle Marchandises, Note Méthodologique, MVA and Kessel + Partner, Paris/Freiburg.

NEA (2000). European transport forecasts 2020, Summary freight transport, report for the European Commission, NEA, Rijswijk.

Newton, S. (2008). Worldnet: applying transport modelling techniques to long distance freight flows. In: *European transport conference*, Noordwijkerhout, The Netherlands.

Nuzzolo, A., & Russo, F. R. (1995). A disaggregate freight modal choice model. In: *Seventh world conference on transport research*, Sydney.

Oum, T. (1979). Derived demand for freight transport and intermodal competition in Canada. *Journal of Transport Economics and Policy*, 13(2), 149–166.

Oum, T. H. (1989). Alternative demand models and their elasticity estimates. *Journal of Transport Economics and Policy*, XXIII(2), 163–188.

Puckett, S. M., & Hensher, D. A. (2006). Modelling interdependent behaviour as a sequentially administrated stated choice experiment: analysis of freight distribution chains. In: *International association of travel behaviour research conference*, Kyoto.

Samimi, A., Mohammadian, A., & Kawamura, K. (2010). Freight demand microsimulation in the U.S. In: *World conference on transport research*: Lisbon.

SCENES consortium. (2001). SCENES transport forecasting model: calibration and forecast scenario results, deliverable 7 to EU DGTREN; SCENES Consortium, Cambridge.

Small, K. (1987). A discrete choice model for ordered alternatives. *Econometrica*, 55(2), 409–424.

Tavasszy, L. A. (1996). Modelling European freight transport flows, PhD. Thesis, Delft University of Technology.

Tavasszy, L. A., Smeenk, B., & Ruijgrok, C. J. (1998). A DSS for modelling logistics chains in freight transport systems analysis. *International Transactions in Operational Research*, *5*(6), 447–459.

Tetraplan (2009). *TENCONNECT*. Copenhagen: Tetraplan A/S.

Train, K. (2003). *Discrete choice methods with simulation*. Cambridge: Cambridge University Press.

van de Kaa, E. J. (2008). *Extended prospect theory, findings on choice behaviour from economics and the behavioural sciences and their relevance for travel behaviour* (Ph.D. thesis). Delft University of Technology.

Wen, C. -H., & Koppelman, F. S. (2001). The generalised nested logit model. *Transportation Research B*, *35*(7), 627–641.

Windisch, E., de Jong, G. C., & van Nes, R. (2010). A disaggregate freight transport model of transport chain choice and shipment size choice. In: *ETC*, Glasgow.

Winston, C. (1981). A disaggregate model of the demand for intercity freight. *Econometrica*, *49*, 981–1006.

7 Vehicle-Trip Estimation Models

José Holguín-Veras[a], Carlos González-Calderón[a],
Iván Sánchez-Díaz[a], Miguel Jaller[a] and Shama Campbell[a]

[a]Center for Infrastructure, Transportation, and the Environment, and the VREF's
Center of Excellence for Sustainable Urban Freight Systems, Department of Civil
and Environmental Engineering, Rensselaer Polytechnic Institute, Troy, NY, USA

7.1 Introduction

Freight demand is widely recognised to be a very complex phenomenon. This is a consequence of its multifaceted and multidimensional nature, the large number of interacting agents and facilities involved (e.g. shippers, carriers, receivers, distribution centres), the role played by the economic linkages between these agents (e.g. independent vs. integrated companies), the diverse functions that are performed (e.g. long haul transport, urban deliveries, parcel service), the existence of multiple modes and vehicle types that could be used to transport and distribute the freight (e.g. airplanes, rail, trucks), the wide range of geographic areas covered (e.g. urban, regional, national), and the multiplicity of approaches to define and measure freight (e.g. tons, vehicle trips). The influence of the latter − central to the estimation of vehicle trips − could be illustrated with the assistance of an example. Figure 7.1 depicts a producer (also serving as carrier) that distributes supplies to three consumers (receivers), from its home base (HB). To transport these goods, the supplier uses a truck to perform a delivery tour with three separate trips (HB to the first stop at S1, where receiver R1 is located), from the first stop to the second stop, S2, (location of receivers R2 and R3), and finally the empty trip back to the HB.

The figure reveals that the start and end of the commodity flows (which marks the physical locations of the production and consumption relations that link the producer to the consumer) are not necessarily the same as those for the vehicle trips (that reflect the path taken by the freight vehicles). The same could be said about the empty trip, which returns to HB from a different direction. This mismatch between the geographic locations of the production−consumption ends, and trip origins and destinations presents a challenge to traffic assignment because not accounting for tour behaviour will lead to incorrect results. This would happen, for instance, if the vehicle trips produced by the transport of supplies from HB to S2 are assigned to the network assuming that they originate in HB, opposed to arriving to S2 from S1.

Modelling Freight Transport. DOI: http://dx.doi.org/10.1016/B978-0-12-410400-6.00007-0

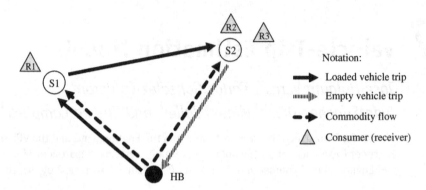

Figure 7.1 Vehicle trips, commodity flows and delivery tours.

The existence of two different units of flow (vehicle trips and commodities) has given rise to two families of freight demand models (vehicle-trip-based and commodity-based models). In a simplified manner, commodity-based models try to predict the commodity flows, typically classified by commodity type, that take place from producers to end/intermediate consumers. The focus on the commodities enables these models to explicitly take into account the economic characteristics and operational constraints associated with the different types of cargo. Once the commodity flows have been estimated, the freight vehicle traffic (both loaded and empty) is computed using vehicle-trip estimation models. In contrast, vehicle-trip-based models attempt to directly forecast the freight traffic generated by economic activity, without going through the intermediate step of estimating the commodity flows. Figure 7.2 shows schematics of the basic model components of these approaches.

Although simpler, the focus on vehicle trips prevents these models from considering: (1) mode/vehicle choice, because they implicitly assume that mode/vehicle choice already took place; (2) the economic characteristics of the cargo and the corresponding operational constraints; and (3) whether or not the vehicles are empty or loaded (Holguín-Veras et al., 2001; Holguín-Veras, Jaller, & Destro, 2010a; Ogden, 1992). The ability of these approaches to represent the freight demand is very different. First, vehicle-trip estimation models cannot explicitly consider the commodity type, which has been found to be a key variable that influence freight behaviour. See for instance Holguín-Veras & Wang (2011) and Holguín-Veras et al. (2013). A second reason is the inability of vehicle-trip-based models to consider mode/vehicle choice, which could be a major drawback if mode choice is a possibility, or if intermodal aspects must be considered by the model. From the standpoint of ability to represent freight behaviour, there shall be no doubt that commodity-based models are better than vehicle-trip-based models. A summary of their features is provided in Table 7.1.

Table 7.1 implies that, in essence, no family of models provides a complete picture of freight markets. Then, it is required the use of complementary models to

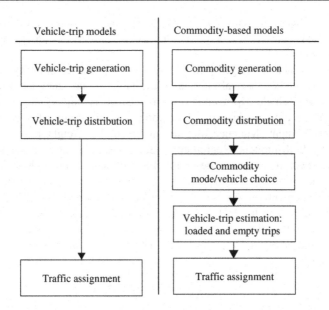

Figure 7.2 Model components of vehicle-trip-based and commodity-based approaches.

Table 7.1 Key Features of Commodity-Based and Trip-Based Models

Aspect	Commodity-Based Models	Vehicle-Trip-Based Models
Ability to consider economic characteristics and/or operational constraints associated with the cargo	High	None
Ability to replicate the real-life process	High	Low
Ability to consider mode choice	High	None
Ability to consider intermodal aspects	High	Low
Need for complementary models	Yes	No
Data requirements	High	Medium−low
Calibration data costs	High−medium	Low

estimate the empty trips. As discussed in Holguín-Veras (2000) and Holguín-Veras & Thorson (2000), this is the result of attempting to force a complex problem, in which both the supply (vehicle trips) and the demand (commodity flows) determine the patterns of freight activity, into a uni-dimensional view that only considers the vehicular supply (as in vehicle-trip-based models) or the demand (as in commodity-based models). Recent developments, such as spatial price equilibrium models that jointly compute commodity flows and vehicle trips/tours, for example, Holguín-Veras, Xu, & Mitchell (2012b), and hybrid models (Donnelly, 2007;

Wisetjindawat, Sano, Matsumoto, & Raothanachonkun, 2007) promise to unify these alternative views. For a comprehensive discussion of tour models, see Holguín Veras et al. (2013).

In spite of the acknowledged limitations, it is clear that commodity-based models are the best methodological alternative available, though they need to be complemented with vehicle-trip estimation models. It is important to stress that vehicle-trip estimation models are as important as the commodity-based models that provide the input data, because the quality of the freight traffic forecast is determined by both models. Regrettably, vehicle-trip estimation models have not received the level of attention deserved. Instead, naïve approaches that do not account for the complexity of underlying dynamics are routinely used. The main objective of this chapter is to raise awareness about appropriate vehicle-trip estimation methodologies, and to identify research needs that ought to be addressed in the future. The chapter discusses the techniques to estimate both loaded trips (Section 7.2) and empty trips (Section 7.3). The final section provides a set of concluding remarks.

7.2   Estimation of Loaded Trips

As its name implies, these models estimate the flow of loaded vehicle trips generated by the transport of commodity flows. Obviously, the first step in the process is to ensure that the commodity flows are converted into a unit suitable for traffic analyses such as tons per day (t/day) or tons per hour (t/hour). This is important because commodity flow models are estimated using aggregated variables such as t/week or even t/year, which are not conducive to transportation network models that require much finer time resolution. The term 'loaded' trip refers to vehicle trips that 'are not empty' as typically freight vehicles carry less than full capacity, i.e. vehicles transporting cargo but with available internal space that could be used to transport more cargo. The estimation of loaded trips is related to choice of mode and the vehicle that would be used. In the case of trucking, the latter is particularly important because the wide range of loading — ranging from 0.50 t for pickup trucks to 30 t and more for large semi-trailers — significantly reduces the validity of using a 'generic truck' to represent the wide range of vehicles. These linkages could be analysed with the assistance of a simplified mathematical model.

Consider an origin–destination (OD) pair i–j where a flow F_{ij} of cargo is transported by different modes/vehicle types r, and the average payloads in that corridor are a_{ijr}. If the probability of selecting vehicle (or mode) r is Q_{ijr} then, the vehicle traffic V_{ij} is equal to:

$$V_{ij} = \sum_r \frac{Q_{ijr} F_{ij}}{a_{ijr}} = \frac{F_{ij}}{\left(\sum_r (Q_{ijr}/a_{ijr})\right)^{-1}} = \frac{F_{ij}}{\overline{a_{ijr}}} \tag{7.1}$$

Assuming that $\overline{a_{ijr}}$ is the mean payload that converts the flow F_{ij} into freight vehicle traffic V_{ij}, it is obvious that $\overline{a_{ijr}}$ is the weighted harmonic mean of the payloads, as shown below (which means that using the arithmetic mean of the payloads would lead to erroneous results):

$$\overline{a_{ijr}} = \left(\sum_r \frac{Q_{ijr}}{a_{ijr}} \right)^{-1} \qquad (7.2)$$

The fact that Eq. (7.2) depends on mode/vehicle choice and the payloads of the alternatives implies that great care must be exercised to ensure that the value of $\overline{a_{ijr}}$ is appropriate because:

- Different OD pairs are likely to have different modal/vehicular splits, due to the nature of the commodities transported, shipment distances, and the availability of transportation alternatives.
- The degree of market competition at an OD pair and industry sector (defined by the commodity type) significantly influences the payloads. In monopolistic conditions, payloads could approach full capacity, because the carriers have the market power to consolidate cargo to the maximum, even if doing so lead to delays to the customers. However, in competitive markets the carriers have very little power to consolidate as the associated delays may lead to their dismissal by irate customers. As a result, load factors (the ratio of payload to capacity) could be as low as 10%.

These considerations highlight the importance of gaining insight into the observed patterns of the payloads by vehicle type. Although the amount of research on this subject is very small, the literature of freight mode/vehicle choice may shed some light. For instance, the vehicle choice models estimated by Holguín-Veras (2002), using disaggregate data from Guatemala indicate that shipment size increases with distance for all the vehicle classes considered: small trucks (pickup trucks), midsize trucks (large 2- and 3-axle rigid trucks), and large trucks (semi-trailers); and that the larger the vehicle the larger the increase in shipment size (Holguín-Veras, 2002, 340). These results make sense as they are consistent with logistic practices and, in particular, with the economic order quantity (EOQ) model (Harris, 1915) that postulates that the larger the transport cost, the larger the shipment size that minimises total logistic cost. These findings are also supported by the findings of Cavalcante & Roorda (2010) who found that shipment size decreases with the value density of the cargo ($/kg), and fuel cost of the vehicle chosen, while it increases with distance (km). The analyses of Holguín-Veras (2002) also revealed that the probability of using semi-trailers increases with distance, while the ones for large trucks and pickup trucks do the opposite. Cavalcante & Roorda (2010) also found that small vehicles are preferred for transport of high value cargo; while larger vehicles are likely to be used for long distances. These patterns directly impact on the generation of loaded trips.

In order to gain insight into how the mean payloads change with trip distance, the authors decided to use the intercity shipment size models of Holguín-Veras

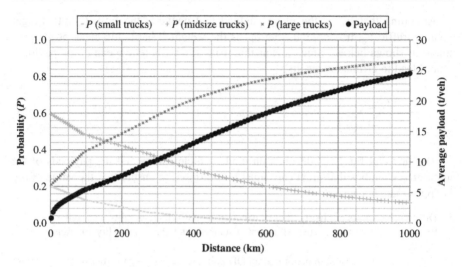

Figure 7.3 Average payload (t/vehicle) and probability of choosing vehicle.

(2002) and assume they provide an estimate of the payloads for a given vehicle class. This is a reasonable assumption, as research has shown that the closer the shipment size is to the capacity of the vehicle, the more likely that vehicle class is selected (Holguín-Veras, 2002). Thus, observed shipment sizes and payloads are correlated. The application of these shipment size models, combined with Eq. (7.2), to the data for Guatemala leads to the results in Figure 7.3. The figure shows that the average payload for the entire freight vehicle traffic stream increases with distance, reflecting the diminishing role of small and midsize trucks in the transport of the cargo. The difference could be substantial. As shown, while for a distance of 2.5 km, the mean payload is 0.81 tons/trip and a total of 1229 vehicle trips are generated (with a significant proportion of small and midsize trucks); for 100 km, the average payload increases to 5.5 t/vehicle requiring 181 vehicle trips; for 500 km, the mean payload increases to 15.67 t/trip producing 64 vehicle trips; and for 1000 km and a mean of payload of 24.5 t/trip the traffic would be of 40 vehicles.

Obviously, these results have notable implications in terms of the amount of externalities produced as research has already established that large trucks produce less externalities, per unit amount of cargo transported, than small trucks (Holguín-Veras, Torres, & Ban, 2011b). Further research is needed to increase the sustainability of the traffic of small vehicles by either inducing shippers to use larger shipments, convincing receivers to accept off-hour deliveries (Holguín-Veras, Silas, Polimeni, & Cruz, 2007, 2008), or fostering the use of cooperative delivery practices that reduce truck traffic.

A final important aspect that is worthy of mention is related to the impacts of tour behaviour on the mismatch between production–consumption and OD of trips illustrated in Figure 7.1. The likely occurrence of such mismatch suggests that the methodologies discussed in this section would work well in cases where the number of delivery stops is relatively small which is typical of intercity travel. Conversely, if long delivery tours are the norm, which is the case of urban areas, the estimates of loaded vehicle traffic are likely to have an error. In such cases, either hybrid models (Donnelly, 2007; Wisetjindawat et al., 2007) or formal freight-tour models (Holguín-Veras et al., 2012b; Holguín-Veras, Thorson, & Mitchell, 2012a; Wang & Holguín-Veras, 2009a, 2009b) offer a better alternative. An issue of concern, however, is that the heuristic procedures typically used to produce the tours in hybrid models tend to lack behavioural support. See Wang & Holguín-Veras (2008), for a discussion. Further research is needed to ensure that the tours estimated by hybrid models are consistent with reality.

7.3   Estimation of Empty Trips

The techniques discussed in the previous section provide estimates of the numbers of loaded vehicle trips. However, they cannot estimate the empty trips, which are a sizeable portion of the total freight traffic representing 20% of the freight traffic in urban areas (Strauss-Wieder, Kang, & Yokei, 1989; Wood, 1970), about 40% of total intercity trips (Holguín-Veras & Thorson, 2003a), and almost 50% of the directional truck traffic in some corridors. Proper estimation of empty trips is important because not doing so could lead to large errors in the estimation of the directional freight traffic, which could be as high as 83% of difference (Holguín-Veras & Thorson, 2003a).

The generation of empty trips is directly influenced by the degree of symmetry of the commodity flow OD matrices and, indirectly, by the type of commodity that generate the loaded trips. The influence of the former is relatively obvious. Taking into account that most freight carriers return to their HBs, it is clear that if the commodity flow matrix is symmetric (the cargo flowing from i to j is equal the one from j to i, for all ODs) then the per cent of empty trips, P_e, could be very low and even zero. At the other end of the spectrum, if all the cargo travels in one direction and nothing in the opposite, the number of empty trips would equal the number of loaded trips and P_e could reach 50%. In real life and in the absence of pathological examples, the higher the degree of asymmetry in the commodity flow matrix, the higher the percentage of empty trips P_e will be. A second factor that indirectly influences the generation of empty trips is the commodity type that is being transported. First, the commodity type captures the degree of asymmetry of the corresponding commodity flow matrix. For instance, a commodity flow OD matrix for vegetables is likely to show large volumes of cargo flowing from production to consumption regions, and not much the other way around. The commodity type also influences the generation of empty trips by means of the technological and operational practices associated with the commodity. For instance, tanker trucks cannot be used to transport general cargo. Similarly, a waste truck may not be allowed to transport food, on account of the health risk.

Although it seems clear that different commodity types generate different amounts of empty trips, the lack of data is a major obstacle. The challenge is that while most OD surveys collect data about commodity flows and loaded trips by commodity type, almost none collect data about what commodity generated the empty trips (only the total number of empty trips are routinely collected). For that reason, the vast majority of empty trip models use as the dependent variable the total number of empty trips. Only the generalised Noortman and van Es (NVE) model, which is discussed later on, is able to consider the commodity type of the cargo being transported.

The estimation of empty trips has been undertaken in a number of different ways, with differing degrees of success. The simplest techniques, and the most problematic, entail the expansion of the loaded trips OD matrix to compensate for the missing empty trips. As loaded trips are computed using mean payload, this implies proportionality between empty trips and commodity flows and it will, thus, lead to major directional errors (Holguín-Veras & Thorson, 2003a). The reason is that, while the loaded trips are proportional to the commodity flow, the empty trips tend to run in the opposite or other directions. This technique, described by Hautzinger (1984) as the 'naïve proportionality model', is very problematic as the empirical evidence shows that empty trips are not correlated with the commodity flows on the same direction (Ozbay, Holguín-Veras, & de Cerreño, 2005). A second technique is to model empty trips as a separate 'commodity' (Fernández, de Cea Ch, & Soto, 2003; Tamin & Willumsen, 1992), which is also questionable because it breaks down the interconnection between loaded and empty trips (the latter is a by-product of the former). Therefore, there is no way to guarantee consistency between empty and loaded trips. More advanced models, the main focus of this section, estimate empty trips as a function of the commodity flows. These models overcome, to a large extent, the limitations of naïve approaches.

7.3.1   Commodity-Based Empty Trip Models

The fundamental principle of these models is that the flow of empty trips can be estimated as a function of commodity flows in the opposite direction. This section discusses the different models that exist.

Notation:

m_{ij} = commodity flow between origin i and destination j in tons
a_{ij} = average payload (tons/trip) for loaded trips between i and j
$x_{ij} = m_{ij}/a_{ij}$ = estimated number of loaded trips between i and j
y_{ij} = estimated number of empty trips between i and j
$z_{ij} = x_{ij} + y_{ij}$ = estimated total number of trips (loaded + empty) between i and j
d_{ij} = distance between origin i and destination j
p = probability of a zero order trip chain
γ, β = parameters to be determined empirically
$P^h(j)$ = probability that a vehicle that came from h to i chooses j as the next destination
$P^h(E/j)$ = probability that a vehicle following the tour $h-i-j$ does not get cargo to j
$P^h(j)P^h(E/j)$ = probability that a vehicle travelling in $h-i-j$ goes empty to j

$P^h(j)$ is a function of the attractiveness of zone j as a destination, which can be assumed to be a function of the commodity flow from j to i, m_{ij} and the trip impedance between i and j.

7.3.1.1   Noortman and Van Es

The first empty trip model that attempted to estimate empty trips using the commodity flows was developed by Noortman & van Es (1978) as part of the Dutch Freight Transport Model (Hautzinger, 1984). In the NVE model the empty trips are estimated as a function of the opposing commodity flow.

$$z_{ij} = x_{ij} + y_{ij} = \frac{m_{ij}}{a_{ij}} + p\frac{m_{ji}}{a_{ij}} = x_{ij} + px_{ji} \qquad (7.3)$$

As shown, the loaded trips are estimated as the commodity flow in this direction (from i to j) divided by the payload; while empty trips are estimated as the commodity flow in the opposing direction (from j to i) divided by the payload and multiplied by a constant p to be determined empirically. The empirical tests conducted indicate that NVE model is fairly good (Holguín-Veras & Thorson, 2003a).

7.3.2   Generalised NVE Models

These formulations estimate empty trips as a function of the cargo flows by commodity type in the opposite direction. This stands in contrast with the standard NVE model that uses the total cargo flow. Mathematically:

$$y_{ij} = \sum_l \beta^l x_{ji}^l \qquad (7.4)$$

where l is an index of commodity type, x_{ji}^l is the number of loaded trips with commodity l travelling from j to i and β^l are parameters to be estimated empirically.

As shown, Eq. (7.4) is a more general version of the NVE model that could be readily estimated using ordinary least squares (OLS). The dependent variable is the total flow of empty trips y_{ij} and the independent variables are the loaded trips, by commodity type, in the opposite direction x_{ji}^l. Since by definition $0 \le \beta^l \le 1$, the estimation process has to ensure that the resulting parameters meet this important constraint. However, quite frequently, due to data inconsistencies, the resulting parameters may violate the expected range. In the case of negative parameters, it is obvious that the data do not support the estimation of a commodity specific parameter. Thus, aggregating those commodity types in a supergroup that includes other similar commodities may increase the quality of the parameters estimated.

Commodity types that produce parameters larger than 1 ($l_{p=1}$) may represent cases where loaded trips are accompanied by an equal number of empty trips, for example, in the transport of commodities that require specialized equipment. In such a case, it is more appropriate to assume that the empty trips generated are equal to the loaded trips in the opposite direction (as shown below):

$$y_{ij}^{l_{p=1}} = x_{ji}^{l_{p=1}} \tag{7.5}$$

Once the commodity types with $p = 1$ have been identified, the empty trips that they generate must be subtracted from the total empty trips y_{ij}, and the OLS procedure rerun for the remainder commodity types. If the resulting parameters both meet the condition $0 \le \beta^l \le 1$ and are statistically significant then the model could be considered final. See Jiménez (2009) for an application of this model.

7.3.2.1 Hautzinger

Hautzinger (1984) shows that the NVE model implies that the difference between the flow of freight vehicles (z_{ij} and z_{ji}) is a function of the difference between the commodity flows (m_{ij} and m_{ji}). According to Hautzinger (1984), this is a problem because the empirical evidence suggests that even in cases of extreme differences in the commodity flows the total vehicle flows tend to be equal. To address this, he formulated a model that considers the base location of the freight vehicles. In his model, the total number of trips in both directions is always equal, and the parameters p_i and p_j denote the probabilities of returning empty for vehicles based in i and j respectively, as shown below:

$$z_{ij} = \frac{(p_i m_{ij} + p_j m_{ji})}{a(1 - (1 - p_i)(1 - p_j))} \tag{7.6}$$

However, the research conducted has not confirmed Hautzinger's hypothesis of symmetry in the total flow of freight vehicles (Holguín-Veras, Sánchez, González-Calderón, Sarmiento, & Thorson, 2011a; Ozbay et al., 2005). Not surprisingly, Hautzinger's model was found to produce large estimation errors (Ozbay et al., 2005).

7.3.3   Tour Based Empty Trip Models

A chief limitation of the empty trip models discussed up to now is that they only consider trips from the HB to a single destination. This is a problem because freight vehicles, particularly in urban areas, undertake delivery tours involving more than one delivery stop. For instance, the number of delivery stops made by freight carriers in the New York City metropolitan area ranges from 3.3 stops/tour (stone and concrete) to 20.0 stops/tour (beverages) (Holguín Veras et al., 2013). There is a need to consider the effects of tour behaviour on the generation of empty trips. In order to incorporate tour behaviour is necessary to introduce the concept of *order of a trip chain model* (Holguín-Veras & Thorson, 2003a), which is defined as 'the number of destinations − in addition to the primary trip − that the model is

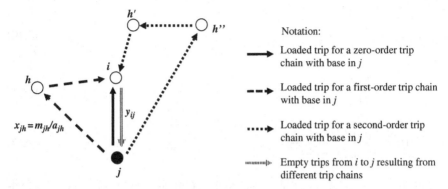

Figure 7.4 Total empty trips generated by trip chains of different orders.

effectively able to consider'. In this context, both NVE and Hautzinger (1984) are zero order trip chain models. Figure 7.4 depicts the loaded and empty trips produced by trip chains of orders zero, one and two to illustrate the rationale of more elaborated empty trips models.

According to Holguín-Veras & Thorson (2003a), since the total number of trips between i and j is not known with certainty, it can be assumed to be a random variable with expected value equal to:

$$E(z_{ij}) = \sum_{n=0}^{N} E(x_{ij}^n) + \sum_{n=0}^{N} E(y_{ij}^n) = \frac{m_{ij}}{a_{ij}} + E(y_{ij}^0) + E(y_{ij}^1) + \sum_{n>1} E(y_{ij}^n) \tag{7.7}$$

As shown, the loaded traffic depends on the commodity and the payload between i and j; while the expected number of empty trips produced by an n order trip chain depends on the likelihood that this trip chain takes place. Eq. (7.7) could be transformed into (Holguín-Veras & Thorson, 2003a):

$$E(z_{ij}) = \frac{m_{ij}}{a_{ij}} + p\frac{m_{ji}}{a_{ji}} + \gamma \sum_{h \neq j} x_{hi}(P^h(j)P^h(E/j)) \tag{7.8}$$

where γ is a parameter to be estimated empirically and the probabilities of choosing a destination can be determined using different specifications.

In general, the destination choice probabilities can be estimated using spatial interaction models. Equations (7.9)–(7.12) show some alternative specifications tested by Holguín-Veras & Thorson (2003a), where the probabilities are a function of the commodity flow or/and the trip impedance.

$$P(j) = \frac{m_{ij}}{\sum_l m_{il}} \tag{7.9}$$

$$P(j) = \frac{m_{ij}e^{-\beta d_{ij}}}{\sum_l m_{il}e^{-\beta d_{il}}} \tag{7.10}$$

$$P(j) = \frac{m_{ij}(d_{ij})^{-\beta}}{\sum_l m_{il}(d_{il})^{-\beta}} \tag{7.11}$$

$$P^h(j) = \frac{m_{ij}(d_{ij}+d_{hi})^{-\beta}}{\sum_l m_{il}(d_{il}+d_{hi})^{-\beta}} \tag{7.12}$$

Equation (7.9) is the most intuitive specification, as it implies that the probability of choosing a destination j is a function of the commodity flow from i to j and the commodity flows to other connecting zones. In Eq. (7.10), the commodity flows are weighted using an exponential function of the corresponding trip impedances; while in Eq. (7.11), the weights are expressed using a power function. In Eq. (7.12), commodity flows are weighted using a power function that accounts for the impedance between i and j, and also for the distance previously travelled. The specification in Eq. (7.12), the most complete one, is the only model that has memory as it takes into account the amount of travel already done. It is also the one that produced the most accurate results (Holguín-Veras & Thorson, 2003a).

Replacing Eqs. (7.9) thru (7.12) into Eq. (7.8), one could obtain:

$$E(z_{ij}) = \frac{m_{ij}}{a_{ij}} + p\frac{m_{ji}}{a_{ji}} + \gamma \sum_{h \neq j} x_{hi} \frac{m_{ij}}{\sum_l m_{il}} P(E/j) \tag{7.13}$$

$$E(z_{ij}) = \frac{m_{ij}}{a_{ij}} + p\frac{m_{ji}}{a_{ji}} + \gamma \sum_{h \neq j} x_{hi} \frac{m_{ij}e^{-\beta(d_{ij})}}{\sum_l m_{il}e^{-\beta(d_{il})}} P(E/j) \tag{7.14}$$

$$E(z_{ij}) = \frac{m_{ij}}{a_{ij}} + p\frac{m_{ji}}{a_{ji}} + \gamma \sum_{h \neq j} x_{hi} \frac{m_{ij}(d_{ij})^{-\beta}}{\sum_l m_{il}(d_{il})^{-\beta}} P(E/j) \tag{7.15}$$

$$E(z_{ij}) = \frac{m_{ij}}{a_{ij}} + p\frac{m_{ji}}{a_{ji}} + \gamma \sum_{h \neq j} x_{hi} \frac{m_{ij}(d_{ij}+d_{hi})^{-\beta}}{\sum_l m_{il}(d_{il}+d_{hi})^{-\beta}} P(E/j) \tag{7.16}$$

The models discussed up to now in this section implicitly assume that the parameter p is constant. However, the empirical evidence and the observations of freight practices suggest that p is variable as it likely depends on trip length, the magnitude of the opposing commodity flow, among others. For instance, carriers may undertake a short trip with an empty backhaul much easier than a comparable long trip, with would lead to a p value that decrease with distance. Holguín-Veras, Thorson, & Zorrilla (2010b) and Ozbay et al. (2005) examined a number of alternative models with p variable. An example of a linear specification that includes the distance is shown in Eqs. (7.17) and (7.18):

$$z_{ij} = x_{ij} + p_0(1 + p_1 d_{ij})x_{ji} \tag{7.17}$$

$$z_{ij} = x_{ij} + p_0 x_{ji} + p_0 p_1 d_{ij} x_{ji} = x_{ij} + p_0 x_{ji} + p_1 w_{Lji} \tag{7.18}$$

Where p_0 and p_1 are parameters to be estimated, w_{Lji} (the product of x_{ji} and d_{ij}) is the vehicle-km travelled while loaded, and the other variables are as defined before. This linear model can be readily estimated using statistical procedures. Furthermore, the inclusion of the w_{Lji} term can be exploited to reflect observed values of veh-km using constrained optimisation techniques.

7.3.4 Parameter Estimation Procedures

This section introduces procedures to obtain the parameters that best replicate empty trip traffic. Two families of techniques are briefly discussed: unconstrained and constrained parameter estimation. The former refers to techniques in which the parameters do not have to meet any constraints; while the latter denotes the cases in which the search constrains one or several parameters to be equal to a preset value.

In the unconstrained parameter search, no external information is incorporated into the estimation procedure, as the parameters are free to take any value that minimises the estimation error. In this context, OLS procedures have been successfully used to calibrate linear empty trip models (Holguín-Veras et al., 2011a). For instance, the optimal parameter of the NVE model that minimises the least squares is (Holguín-Veras & Thorson, 2003b):

$$p^* = \frac{\sum_{i,j} y_{ij} x_{ji}^e}{\sum_{i,j} (x_{ji}^e)^2} \tag{7.19}$$

where x_{ji}^e is the estimated loaded trip matrix, and y_{ij} is the actual empty trip matrix.

However, since these methods focus exclusively on finding the lowest estimation errors, the parameter estimates may not necessarily be consistent with external sources of information, for example, area wide estimates of the per cent of empty trips. This could be problematic.

Constrained parameter search tries to ensure that the model replicates the observed percentage of empty trips, p_e. This parameter is rather stable across the world, as it is frequently reported to be between 20−30% in urban areas (Ogden, 1992; Strauss-Wieder et al., 1989), and about 30−50% in intercity transport (Holguín-Veras & Thorson, 2003a). The stability of p_e is important because the parameters of the empty trip models implicitly determine the number of empty trips produced and p_e. Thus, a good empty trip model should reproduce the per cent of empty trips observed in real life. The interconnection between empty trip models and p_e can be illustrated with the derivations based on the NVE model:

$$z_{ij} = x_{ij} + y_{ij} = \frac{m_{ij}}{a} + p\frac{m_{ji}}{a} = x_{ij} + px_{ji} \tag{7.20}$$

The total number of trips in the model is:

$$\sum_{i,j} z_{ij} = \sum_{i,j} x_{ij} + p\sum_{i,j} x_{ji} \tag{7.21}$$

If Z^* represents the total number of trips, X^* the total loaded trips, and Y^* the total empty trips in the region, Eq. (7.21) could be written as:

$$Z^* = X^* + Y^* = X^* + pX^* \tag{7.22}$$

Then, the percentage of empty trips, p_e is:

$$p_e = \frac{Y^*}{Z^*} = \frac{Y^*}{X^* + Y^*} = \frac{pX^*}{X^* + pX^*} = \frac{p}{1+p} \tag{7.23}$$

Thus, the parameter p of the NVE model can be calculated as:

$$p = \frac{p_e}{1 - p_e} \tag{7.24}$$

Equations (7.23) and (7.24) show the interrelation between p and p_e. If $p = 0$, no empty trips are generated and $p_e = 0$. At the other end of the spectrum, if $p = 1$ (the maximum value possible because it is a probability), then $p_e = 0.5$. The latter corresponds to the case in which the cargo only flows in one direction. These results imply that if one knows the per cent of empty trips, p_e, with reasonable precision, one could estimate the parameter p without having to undertake a formal process of parameter estimation. The work of Holguín-Veras & Thorson (2003b) suggests that this process works quite well. Since producing an estimate of p_e is certainly easier than collecting and OD matrix of empty trips, this insight could simplify data collection procedures.

Other forms of constraints could also be used. For instance, if one considers the empty trips as a function of trip length, taking into account the vehicle miles travelled for loaded trips, the total trips between zones i and j will be:

$$z_{ij} = x_{ij} + p_0x_{ji} + p_1w_{Lji} \tag{7.25}$$

If Z^* represents the total number of trips, X^* the total loaded trips, Y^* the total empty trips in the region, and W_E^* and W_L^* represents the empty and total vehicle miles travelled by loaded trucks respectively, Eq. (7.25) could be written as:

$$Z^* = X^* + Y^* \tag{7.26}$$

$$Z^* = X^* + p_0 X^* + p_1 W_L^* \tag{7.27}$$

Then, the percentage of empty trips p_e is:

$$p_e = \frac{Y^*}{Z^*} = \frac{Y^*}{X^* + Y^*} = \frac{p_0 X^* + p_1 W_L^*}{X^* + p_0 X^* + p_1 W_L^*} \tag{7.28}$$

Since $p = (p_e / 1 + p_e)$, it is easy to show that

$$X^* = \frac{p_0 X^*}{p} + \frac{p_1 W_L^*}{p} \tag{7.29}$$

Ozbay et al. (2005) showed that the constrained parameters p_0 and p_1 are a function of the parameter p, and are given by:

$$p = \frac{p_e}{1 + p_e} \tag{7.30}$$

$$p_0 = \frac{p X^* S^* - W_L^* W_E^*}{X^* S^* - (W_L^*)^2} \tag{7.31}$$

$$p_1 = \frac{(W_E^* - p W_L^*) X^*}{X^* S^* - (W_L^*)^2} \tag{7.32}$$

This shows that as the parameter p increases, the parameter that captures the relationship between empty trips and opposing commodity flows, p_0, increases too. However, as p increases, the parameter that captures the contribution of vehicle miles to empty trips, p_1, reduces.

Similar derivations could be conducted for the tour models. If the probabilities $P(E/j)$ are assumed to be constant and equal to $P(E/\bullet)$ one could obtain (Holguín-Veras & Thorson, 2003a):

$$p_e = \frac{Y^*}{Z^*} = \frac{Y^*}{X^* + Y^*} = \frac{p X^* + \gamma X^* P(E/\bullet)}{X^* + p X^* + \gamma X^* P(E/\bullet)} = \frac{p + \gamma P(E/\bullet)}{1 + p + \gamma P(E/\bullet)} \tag{7.33}$$

Equation (7.33) could be used to ensure that the parameters p and γ reproduce the observed value of the p_e. This will result into better estimates of both empty and total freight traffic in the study area.

7.3.5   Empirical Evidence

A number of studies (González-Calderón, Sánchez-Díaz, Holguín-Veras, & Thorson, 2011; Holguín-Veras & Thorson, 2003a; Holguín-Veras et al., 2010b; Holguín-Veras et al., 2011a; Jiménez, 2009) have used different data sets (e.g. Dominican Republic, Guatemala, Colombia) to test the models discussed in the chapter. The empirical evidence unambiguously confirms the importance of modelling empty trips, as not doing so increase the root mean squared errors of the total flow by 57% for intercity movements and by 83% for suburban movements (Holguín-Veras & Thorson, 2003b). These analyses also show that p_e is typically around 30%. For instance the analyses of seven national freight OD surveys made in Colombia from 2000 to 2005 indicate that p_e ranged from 26.4% to 30.0% (Holguín-Veras et al., 2011a). However, the models estimated in Holguín-Veras et al. (2011a) and González-Calderón et al. (2011) using data from Colombia found that p_e is slightly time dependent as it has changed over the years.

Jiménez (2009) developed three different econometric empty trips models using Colombian data: (1) an aggregate model; (2) a disaggregated model by vehicle type; and (3) a disaggregated model by vehicle and by commodity. The author found that the disaggregated models by vehicle and by commodity replicate the trips of the databases better than the other models.

The performance of the various models discussed in this chapter has been assessed with OD data. Holguín-Veras and Thorson (2003b) applied both the NVE and the tour models to data from Guatemala and found that, as expected, that the unconstrained models performed better than constrained models, though by less than 5%. In all cases, the tour models outperformed the NVE model. In situations where delivery tours are the norm (urban trips), the best tour model outperformed NVE model by about 9%. In contrast, when applied to the intercity data where delivery tours are less frequent, the best tour model outperformed NVE by only 1.3% (unconstrained) and 5% (constrained). This makes sense as the tour models are a more general version of NVE.

González-Calderón et al. (2011) studied the impacts of spatial and temporal aggregation of empty trips models using data from Colombia. They used three different aggregation levels (28, 36 and 69 transportation analysis zones, TAZs) to analyse five empty trips models: the NVE model and the tour models of Holguín-Veras & Thorson (2003b). They found that the use of large TAZs masks the underlying tour behaviour, and the NVE model produces better results than the tour models. However, the tour models perform much better if smaller TAZs are used because the TAZs can capture tour behaviour patterns. In terms of temporal aggregation, they found, as Holguín-Veras et al. (2011a), that the time dependence is minimal for the p parameters over time for all empty trips models.

Holguín-Veras et al. (2011a) analysed the temporal effects on parameter of freight demand models based on cross-sectional and panel data. They examined the stability of parameters in trip generation, trip distribution, and empty trips using the Colombian national freight OD surveys and used OLS to estimate NVE models for the various years. The cross-sectional model resulted in values of p ranging from

0.369 to 0.444. The panel model included a term to capture fixed time effects for each year and a time-dependent parameter. The analysis revealed that the parameter p slightly decreases over the years, though the rate of change is quite small (about 0.83% per year), which is a rate of change slower than the one for the corresponding the trip generation and distribution models.

7.4 Concluding Remarks

This chapter has provided an overview of the vehicle-trip estimation techniques that could be used to quantify the amount of freight traffic, both loaded and empty, that would be generated by transporting commodity flows in the study area. The chapter establishes that the estimation of loaded traffic must account for the underlying patterns of mode and vehicle choice, their influence on shipment sizes and payloads and tour behaviour. The numerical estimates produced, based on the shipment size models estimated by Holguín-Veras (2002) make clear that estimating loaded traffic on the basis of a mean payload that applies to all ODs is likely to lead to erroneous results. The reason is that the mean payloads are likely to change as a function of shipment distance. Since these changes are most significant for the short range of distances suggests that care must be exercised when estimating vehicle trips in urban areas.

A second aspect of importance is the impact of tour behaviour on traffic assignment. The standard practice of pickup/delivery tours leads to a situation in which the locations of the trip ODs do not match the locations that mark the production—consumption relations linking shippers to receivers. In general, the assumption that the freight traffic flows in the same direction as the commodity flows is only valid if tours are the exception and not the norm. Potential ways to address this problem include the use of hybrid and formal tour models (Donnelly, 2007; Holguín-Veras et al., 2012b; Wang & Holguín-Veras, 2009a; Wisetjindawat et al., 2007). Both families of approaches need further improvements. Research is still needed to bolster the behavioural assumptions used in hybrid models to ensure that they correctly capture the behaviours observed in real life. Formal tour models, though showing great promise, still need further development and testing to ensure they provide reasonable results in real-life applications.

The chapter also provides an overview of the methodologies that could be used to estimate empty trips using commodity flows as the input. The chapter discusses empirical evidence that clearly shows that the lack of properly modelling empty trips lead to significant errors in the estimates of directional freight traffic. These estimates indicate that not modelling empty trips increases the root mean squared error of the estimation between 57% and 83% with respect to the results provided by the best empty trip model (Holguín-Veras & Thorson, 2003a). Fortunately, there are a number of methodologies that have been found to produce fairly good estimates of empty trip traffic, and are vastly superior to any of the naïve approaches that have been used (e.g. expanding the loaded trip OD matrix to compensate for the missing empty trips, modelling empty trips as a 'commodity'). The use of such naïve techniques is not advised because it will likely lead to estimation errors.

The simplest conceptually valid model is the one developed by Noortman & van Es (1978). This NVE model, the foundation for all the models that followed, assumes that the empty trip traffic in any given direction is a function of the loaded traffic in the opposite direction. Its simple linear form facilitates its incorporation into commodity-based models. Holguín-Veras & Thorson (2003b) developed a formula to compute the parameter of the NVE model without needing to use a formal optimisation procedure. Moreover, if there are no OD matrices available to compute the optimal parameter but a reasonable guess of the per cent of empty trips could be produced, then the parameter of the NVE model could be estimated using the derivation of Holguín-Veras & Thorson (2003a) that links the per cent of empty trips, P_e, and the parameter p of the NVE model. Another suitable alternative is the generalised NVE model, which estimates the empty traffic as a linear combination of the contributions of the various commodity types. However, none of these models consider tour behaviour, which is a major problem as pickup/delivery tours are frequent.

The state of the art in empty trip modelling is defined by the formulations that account for tour behaviour. These models postulate that the empty traffic for a given OD pair is the summation of the empty trips produced by tours of different lengths. Using probability and spatial interaction principles, a number of mathematical models have been developed (Holguín-Veras & Thorson, 2003a; Holguín-Veras et al., 2010b; Ozbay et al., 2005). The empirical applications conducted clearly establish that the tour models outperform the NVE models if freight tours are the norm. In cases where tours are rare, the performance of tour models and NVE is practically the same. For example, González-Calderón et al. (2011) tested these empty trips models considering tours in Colombia for six consecutive years. The authors confirmed that when considering trip chains, the empty trips models outperform the NVE models for all years with a lower error in all cases.

This chapter provides a comprehensive overview of the methodological landscape concerning vehicle-trip estimation. The authors hope that these suggestions and analyses help both the practitioner and research communities to develop and implement new and enhanced freight demand models.

References

Cavalcante, R., & Roorda, M. (2010). A disaggregate urban shipment size/vehicle-type choice model (TRB 10-3878). In: *Annual meeting of the transportation research board*, Washington, DC.

Donnelly, R. (2007). A hybrid microsimulation model of freight flows. In E. Taniguchi, & R. G. Thompson (Eds.), *Proceedings of the fourth international conference on city logistics* (pp. 235–246). Crete, Greece: Institute for City Logistics11–13 July 2007.

Fernández, L. J. E., de Cea Ch, J., & Soto, A. (2003). A multi-modal supply-demand equilibrium model for predicting intercity freight flows. *Transportation Research Part B: Methodological, 37*(7), 615–640.

González-Calderón, C. A., Sánchez-Díaz, I., Holguín-Veras, J., & Thorson, E. (2011). An empirical investigation on the impacts of spatial and temporal aggregation of empty trips models. In: *4th national urban freight conference*, Long Beach CA, METRANS.

Harris, F. (1915). *Operations and costs*. Chicago, IL: A.W Shaw.

Hautzinger, H. (1984). The prediction of interregional goods vehicle flows — some new modeling concepts. In: *Proceedings of the 9th international symposium on transportation and traffic theory*, Delft, the Netherlands.

Holguín-Veras, J. (2000). A framework for an integrative freight market simulation. In: *IEEE 3rd annual intelligent transportation systems conference*, Dearborn, MI, IEEE.

Holguín-Veras, J. (2002). Revealed preference analysis of commercial vehicle choice process. *Journal of Transportation Engineering*, *128*(4), 336.

Holguín-Veras, J., Jaller, M., & Destro, L. (2010a). Feasibility study for freight data collection — Final report, from <http://www.utrc2.org/research/assets/190/NYMTC-Freight-Data-Final-rpt1.pdf>.

Holguín-Veras, J., List, G., Meyburg, A., Ozbay, K., Paaswell, R., Teng, H., et al. (2001). An assessment of methodological alternatives for a regional freight model in the NYMTC region, from <http://www.nymtc.org/data_services/freight_model/files/Final_Report_053001.PDF>.

Holguín-Veras, J., Sánchez, I., González-Calderón, C., Sarmiento, I., & Thorson, E. (2011a). Time-dependent effects on parameters of freight demand models. *Transportation Research Record*, *2224*(1), 42−50. Available from http://dx.doi.org/10.3141/2224-06.

Holguín-Veras, J., Silas, M. A., Polimeni, J., & Cruz, B. (2007). An investigation on the effectiveness of joint receiver−carrier policies to increase truck traffic in the off-peak hours: Part I: The behaviors of receivers. *Networks and Spatial Economics*, *7*(3), 277−295. Available from http://dx.doi.org/10.1007/s11067-006-9002-7.

Holguín-Veras, J., Silas, M. A., Polimeni, J., & Cruz, B. (2008). An investigation on the effectiveness of joint receiver−carrier policies to increase truck traffic in the off-peak hours: Part II: The behaviors of carriers. *Networks and Spatial Economics*, *8*(4), 327−354. Available from http://dx.doi.org/10.1007/s11067-006-9011-6.

Holguín-Veras, J., & Thorson, E. (2000). An investigation of the relationships between the trip length distributions in commodity-based and trip-based freight demand modeling. *Transportation Research Record*, *1707*, 37−48.

Holguín-Veras, J., & Thorson, E. (2003a). Modeling commercial vehicle empty trips with a first order trip chain model. *Transportation Research Part B*, *37*(2), 129−148.

Holguín-Veras, J., & Thorson, E. (2003b). Practical implications of modeling commercial vehicle empty trips. *Transportation Research Record*, *1833*, 87−94.

Holguín-Veras, J., Thorson, E., & Mitchell, J. (2012a). Spatial price equilibrium model of independent shipper−carrier operations with explicit consideration of trip chains (in review).

Holguín-Veras, J., Thorson, E., & Zorrilla, J. C. (2010b). Commercial vehicle empty trip models with variable zero order empty trip probabilities. *Networks and Spatial Economics*, *10*, 241−259. < http://dx.doi.org/10.1007/s11067-008-9066-7 >.

Holguín Veras, J., Thorson, E., Wang, Q., Xu, N., González-Calderón, C., Sánchez-Díaz, I., & Mitchell, J. (2013). Urban Freight Tour Models: State of the Art and Practice. In M. Ben-Akiva, E. Van de Voorde & H. Meersman (Eds.), Freight Transport Modelling (pp. 335−351). Bingley, UK: Emerald Group.

Holguín-Veras, J., Torres, C., & Ban, J. (2011b). On the comparative performance of urban delivery vehicle classes. *Transportmetrica*, *9*, 1−24. Available from http://dx.doi.org/10.1080/18128602.201.523029.

Holguín-Veras, J., & Wang, Q. (2011). Behavioral investigation on the factors that determine adoption of an electronic toll collection system: Freight carriers.. *Transportation Research Part C, 19*(4), 593−605.

Holguín-Veras, J., C., Wang, S. D., Hodge, Campbell, S., Rothbard, S., Jaller, M., et al. (2013). Unassisted off-hour deliveries and their role in urban freight demand management (in review).

Holguín-Veras, J., Xu, N., & Mitchell, J. (2012b). A dynamic spatial price equilibrium model of integrated production-transportation operations considering freight tours (in review).

Jiménez, A. E. (2009). *Calibración de un modelo para estimar los viajes de carga vacíos−Caso colombiano* (MS thesis). Universidad Nacional de Colombia.

Noortman, H. J., & van Es, J. (1978). Traffic model. Manuscript for the Dutch freight transport model. The Netherlands.

Ogden, K. W. (1992). *Urban goods movement: A guide to policy and planning*. Brookfield, VT: Ashgate Publishing Company.

Ozbay, K., Holguín-Veras, J., & de Cerreño, A. (2005). Evaluation study of New Jersey turnpike authority's time of day pricing initiative, from <http://www.cait.rutgers.edu/final-reports/FHWA-NJ-2005-012.pdf>.

Strauss-Wieder, A., Kang, K., & Yokei, M. (1989). The truck commodity survey in the New York−New Jersey metropolitan area. In: *Good transportation in urban areas. Proceedings of the 5th conference*, Santa Barbara, CA.

Tamin, O. Z., & Willumsen, L. G. (1992). Freight demand model estimation from traffic counts. In: *PTRC annual meeting*, England, University of Bath.

Wang, Q., & Holguín-Veras, J. (2008). Investigation of attributes determining trip chaining behavior in hybrid microsimulation urban freight models. *Transportation Research Record, 2066*, 1−8.

Wang, Q., & Holguín-Veras, J. (2009a). Tour-based entropy maximization formulations of urban commercial vehicle movements. In: *International symposium on transportation and traffic theory*. Hong Kong, China (CDROM).

Wang, Q., & Holguín-Veras, J. (2009b). Tour-based entropy maximization formulations of urban commercial vehicle movements. In: *2009 annual meeting of the transportation research board* (CDROM), Washington, DC.

Wisetjindawat, W., Sano, K., Matsumoto, S., & Raothanachonkun, P. (2007). Microsimulation model for modeling freight agents interactions in urban freight movement. In: *86th annual meeting of the transportation research board*, Washington, DC.

Wood, R. T. (1970). Measuring freight in the tri-state region. In: *The urban movement of goods* (pp. 61−82). Paris: OECD.

8 Urban Freight Models

Antonio Comi[a], Rick Donnelly[b] and Francesco Russo[c]

[a]Department of Enterprise Engineering, University of Rome "Tor Vergata", Rome, Italy
[b]Parsons Brinckerhoff, Inc., Albuquerque, NM, USA
[c]Facoltà di Ingegneria, Università Mediterranea di Reggio Calabria, Reggio Calabria (RC), Italy

8.1 Introduction

Urban freight transport (UFT) is a fundamental component of city life. Every day, people consume and use goods (e.g. food, clothes, furniture, books, cars and computers) produced throughout the world. Furthermore, freight transport maintains a set of core relationships within urban areas since a city is an entity where production, distribution and consumption activities are located and use limited land. Therefore, UFT plays an essential role in satisfying the needs of citizens but, at the same time, produces significant impacts on city sustainability. Sustainability is intended in the economic, social and environmental sense.

Analyses of UFT traditionally focus only on restocking flows, that is freight vehicle flows from warehouse/distribution centres to trade or service establishments (e.g. shops, food-and-drink outlets, service activities), and usually neglect shopping flows. But, according to some statistics collected in European cities (Gonzalez-Feliu, Ambrosini, Pluvinet, Toilier, & Routhier, 2012), the shopping flows (i.e. end-consumers' movements) represent between 45 and 55% of the total goods traffic. These flows are usually not included in freight transport, but counted as passenger transport movements. Besides, end-consumer choices in relation to type of purchasing undoubtedly impact on freight distribution flows: the characteristics of the restocking process are strictly related to the type of retail activities to be restocked in terms of delivery size, delivery frequency, freight vehicle type and so on. Furthermore, end-consumer shopping choices depend on the commercial supply with respect to residence and on end-consumer behaviour, which in turn depends on some characteristics, such as age, income, family dimension and lifestyle. Based on these statements, the chapter discusses specific challenges and applications for urban distribution.

Urban freight transport and logistics are mainly related to the last miles of supply chains, and the companies' strategies have to be compared to the collective interests related to urban freight transportation and logistics. Around the world, several city administrators are looking at UFT issues and in order to reduce its

Modelling Freight Transport. DOI: http://dx.doi.org/10.1016/B978-0-12-410400-6.00008-2

adverse impacts, different city logistics measures have been implemented (Lindholm, 2013; Russo & Comi, 2011). Then, we need to verify if the feasible measures strongly penalise the city centres and the commercial activities located therein (e.g. reducing the city centre accessibility both for users and freight operators). Since the characteristics of urban areas can differ substantially, city logistics measures have to be specifically designed and assessed in order to implement the most effective. In this process, a key role is played by demand models as they can be used to assess the effects of the scenario to be implemented. The models for scenario assessment thus have to investigate the variables that can play an important role in successful scenario implementation. For example, while all measures may be expected to perform well in terms of reducing the external costs of transport, some might increase internal transport costs incurred by some freight actors (e.g. carriers and hence wholesalers). Turnquist (2006) argues that models should be relevant to needs of decision makers, include important behaviour and interactions, supported by data, and verifiable and understandable. It follows that different approaches might be suitable in different circumstances. If just understanding the level of truck flows within urban areas and their impact upon network levels of service is required then some of the models described earlier might suffice. However, in most arenas the need for analytical and behavioural rigor goes much deeper, requiring a more flexible and holistic modelling framework. Therefore, in recent years researchers have sought to develop models that support city logistics. In this paper, an overview of such models is presented.

Starting from authors' literature review (Anand, Quak, Van Duin, & Tavasszy, 2012; Donnelly, 2009; Russo, 2013, chap. 18; Russo & Comi, 2010), the general objective of this chapter is to model the urban freight movements, mapping the behaviour of the retailers and some aspects of the end consumers that generates freight movements in an urban context. To organise a reader framework, we distinguish two macrosegments in the last miles of the freight supply chain, with the retail outlet as final decoupling point: the segment upstream, between firms, and the segment downstream, between consumer and retailer.

Regarding the end consumer, although there are many alternatives, the macrobehaviour may be summarised into two classes (Figure 8.1):

- pull-type behaviour, the end consumer arrives at the purchasing place (e.g. zone d), performs the transaction and purchases the commodity; the user transports the good to the consumption site (e.g. zone o); both in going from o to d and from d to o, the user may make other stops;
- push-type behaviour, the end consumer may or may not go to the purchasing place (e.g. zone d), perform the transaction and purchase the good; the commodity is transported to the site of consumption (e.g. zone o) by actors other than the user.

As for the end consumer, the retailer's macrobehaviour may be also summarised into two classes (Figure 8.1):

- pull-type behaviour, the retailer goes to the acquisition place (e.g. internal zone w or external zone z), purchases (acquires) the goods; the retailer transports the goods to the retail outlet d; along the path the retailer may undertake other stops;
- push-type behaviour, the retailer may or may not go to the place (e.g. internal zone w or external zone z), purchases (acquires) the goods; the goods are transported to sales outlet d by actors other than the retailer.

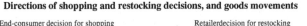

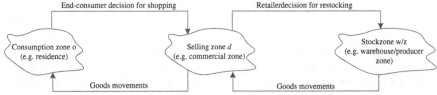

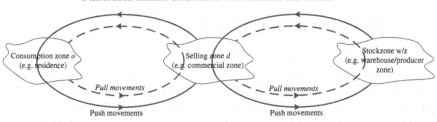

Figure 8.1 Macro-behaviours of end consumers and retailers.

Besides, although the ho.re.ca. (hotel, restaurant, catering) activities and final business consumers (including craftsmen) are end consumers because of being the final destination of goods, we assimilate them to retailers. In fact, their decisional process is quite similar to retailers because of large quantity moved and the distribution channel used.

The end-consumer pull-type movements were prevalent until the arrival of e-shopping, while the first examples of push ones refer to door-to-door selling or the travelling sales of goods. From the retailer's standpoint, pull and push movements are variously interwoven and both may be found in the same decision maker in the restocking phases. Hence in the same shop and with the same decision maker, there may be classes of goods that are restocked with pull movements and others that are restocked with push movements. Furthermore, as synthetised in Figure 8.1, in pull-type movements, both end consumer and retailer travel with purchases and his/her decisional process is mainly related to purchases (i.e. the selling zone d or the stock w/z zone is generally chosen according to the products to buy), while in the push-type ones the attention is on trip and hence the decisional process refers to trip choices (e.g. the stock zone w/z is chosen according to the distribution costs due to retailer to restock).

Note that the push movements are predominant in urban areas (as demonstrated some surveys carried out in some cities, Ambrosini & Routhier, 2004) and are those mainly related to vehicle issues. Besides, while a long time ago, the push movements to end consumers were mainly related to home deliveries by retailers (i.e. a lot of the daily goods were delivered at home by retailers), in the recent years, with the rise of internet, the issues related to push movements of end consumers are growing. The retailers (or better the sellers) deliver large goods, such as furniture and large electronic goods, as televisions or dish washers (Visser, Nemoto, & Browne, 2013). Then, the decision makers for the consignment (e.g. the sender) are organising themselves following the same process of retailer restocking. Therefore, in Section 8.2, due to their homogeneity, the push models for retailer and end consumer will be reviewed. Sequentially, Section 8.3 focuses on pull models both for retailer and end consumer. Finally, Section 8.4 summarises the chapter and provides areas for further researches.

8.2 Push Models of Urban Freight

The 'push' side of urban freight demand represents the collective actions of firms and importers, which compliment the 'pull' consumption dynamics exerted by businesses and consumers. In this section, many of the traditional and emerging factors and modelling approaches influencing the 'push' side of urban freight are described. These include earlier urban truck models, which push flows through the transport system as a function of employment or land use. Extensions to this practice to better represent tours have been advanced. However, a number of economic, supply chain and firmographic advances have changed the landscape of urban freight modelling. Modelling approaches best suited to understanding and predicting these important dynamics are discussed within the framework of how businesses push goods through networks and markets to meet the intermediate and final needs of businesses and households.

8.2.1 Classical Urban Truck Models

The simplest, and perhaps most enduring, example of this type of model is the classical urban truck model. Truck trips are generated as a function of employment, and distributed and routed to destinations using the traditional four-step trip modelling paradigm (five-step if we add the temporal dimension) used for person travel forecasting in urban areas. This process is illustrated in Figure 8.2. The production, consumption and exchange of goods by firms and households within the urban area are represented in abstract terms in the generation and distribution steps. Trips are typically generated for different types of trucks (e.g. light, medium and heavy trucks). This obviates the need for the mode choice step, as other modes of transport are thought to be rarely practical in the urban context. The routing of the trucks along least cost paths through an abstract representation of the urban street system is carried out during network assignment.

The practice of modelling trucks in this manner goes back at least 40 years, with Wigan (1971) and Maejima (1979) describing the derivation and application of such models in London. Schlappi, Marshall, & Itamura (1993) developed an extension based upon survey data from the San Francisco Bay Area that explicitly accounted for garaged trucks in an attempt to capture the differences between

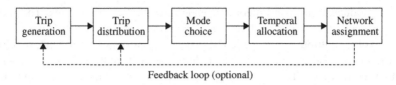

Feedback loop (optional)

Figure 8.2 Sequential trip-based modelling paradigm.

privately owned and for-hire trucks. They also incorporated a special generator model for truck traffic to and from marine ports in the region. Ruiter (1992) built a similar model for Phoenix based upon commercial vehicle survey data. The Quick Response Freight Manual (QRFM) (Cambridge Systematics, 2007) was developed under federal funding in the United States to codify best practices in trip-based urban truck modelling and was largely based on the Phoenix model. A large number of urban truck models have since been implemented using the QRFM approach. Kuzmyak (2008) describes several in the United States that well illustrate the state of practice.

Such models have also been applied in reverse, in the form of synthetic matrix estimation (SME). Such models attempt to adjust an estimated, obsolete or partially observed trip matrix to match observed truck counts. Muñuzuri, Larrañeta, Onieva, & Cortés (2004) developed an SME model for truck movements in Seville that included five different retail markets and one for home deliveries. The demand was consolidated into a single seed matrix and adjusted using a gradient descent method developed by Spiess (1987). More recent formulations have admitted multiple sources of data with reliability estimates attached to each, ability to handle multiple classes of vehicles and use of linear programming techniques to reduce untoward responses to small changes in the traffic count inputs. The model developed by List, Konieczny, Durnford, & Papayanoulis (2001) for New York City remains one of the most innovative synthetic truck models in use, while Holguín-Veras & Patil (2008) advance a multicommodity variant that explicitly accounts for empty truck trips.

While arguably appropriate for representing person travel, the application of the sequential modelling process for urban freight has been widely criticised. The motivation for and characteristics of person travel are well informed by an extensive body of survey research and can be efficiently represented by relatively few market segments, homogeneous household and travel characteristics, and similar travel budgets. Most person travel is characterised by round trips from home to principal destination and back again. Stops are sometimes made along the way for secondary purposes, which the traveller does to simultaneously increase their utility while minimising travel cost. Recent advances in person travel modelling have focused on the explicit representation of person tours or activity chains to better represent this understanding of travel behaviour (Davidson, Donnelly, & Vovsha, 2007).

Unfortunately, freight does not emanate from or move according to the same principles as person flows. Freight flows are, 'the economy in motion', the trade between producers and consumers that underpins modern economies. The factors driving the economy are more diverse and complex than those motivating personal travel, involve multiple entities (such as producers, carriers, distributors, regulators and consumers) and are optimised to reduce the cost and uncertainty associated with their conveyance. Trucks are far less likely to make round trips serving only a single customer per trip, as the lower productivity compared to trip chaining would be prohibitive for many firms. Moreover, the widespread adoption of just-in-time and supply chain logistics has increased the use of distribution centres and transshipment terminals. None of these dynamics can be explicitly represented within the sequential trip modelling process. Not surprisingly, such models do not replicate observed conditions very well and have an uneven record at best in implementation (Taylor & Button, 1999; Wigan & Southworth, 2006).

A number of alternative approaches to the four-step modelling paradigm have been explored in order to overcome these limitations. Some have followed the trend in person travel demand modelling towards activity-based formulations, while others have moved far from it with systems that better represent the diversity of decision makers and dynamics inherent in urban freight. The majority of this chapter is devoted to such approaches. Innovative approaches from the push perspective are discussed in the following sections, while those from the customer's perspective (pull approaches) are addressed in Section 8.3. The emerging modelling approaches described in Section 8.4 are capable of treating both perspectives in a holistic manner.

8.2.2 Tour-Based Extensions to Classical Models

Had trip-based models proven adequate for portraying the overall level of truck traffic within urban areas they would still have fallen from favour. Even if the origin−destination (O−D) patterns were correct, the practice of routing each O−D interchange separately results in flow patterns that do not match observed conditions (Donnelly, 2006). The efficiencies gained by chaining trips together into multistop tours have been long acknowledged by both transport and supply chain modellers. Slavin (1979) advanced the first known truck tour model, based upon survey data collected in Boston. Tours were generated as a function of employment, as in trip-based modelling, as well as accessibility measures, vehicle supply and observed degree of tour formation by industry. A destination choice model was adopted, as gravity models of trip distribution proved unsatisfactory. Southworth (1982) reported a similar finding when modelling truck stops from tour data from Chicago.

More recent research and modelling work have shed considerable light on the topic. Hunt, Stefan, & Brownlee (2006) and Hunt & Stefan (2007) used a commodity flow survey of 3454 business establishments in Calgary conducted in 2000 to build a tour-based commercial vehicle model. They found that freight movements only constituted a third of all commercial vehicle trips. Their survey remains the largest known urban commodity flow survey in existence.

Holguín-Veras & Patil (2005) report on a similar survey in Denver conducted in 1998–99, where approximately 4600 firms returned surveys on 502 vehicles that operated on the survey day. The paper provides an in-depth view of truck tour characteristics that is singular in the literature. The only drawback of their analysis is the lack of information about how tours vary by commodity classification, as that attribute was regrettably not collected.

Figliozzi (2007) developed an idealised continuous approximation model of tour generation for distribution centres, seeking to minimise total kilometres of travel required. Four types of tours were modelled, depending upon the binding constraint (truck capacity, frequency of delivery, tour duration and time windows). One interesting finding was that the percentage of empty trips does not influence the overall efficiency of the generated tours, which flies in the face of conventional wisdom that empty trips are symptomatic of suboptimal and inefficient resource allocation. An equally interesting, but quite different contribution was advanced by Figliozzi, Kingdon, & Wilkitzki (2007). They tracked the routing of a single truck driver in Sydney over an 8-month period. During that time the driver visited 190 different establishments, although almost three quarters of all deliveries were made to only 20% of them. Despite that it was found that over the 8 months that no exact tour (ordered itinerary of stops at specific firms) was repeated.

Holguín-Veras & Thorson (2003) implemented a tour-based model of empty commercial vehicles in Guatemala City that was linked to previous trips in the tour. Their contribution is significant, in that it is one of only a few that explicitly addresses empty vehicle movements, which account for between 20 and 30% of urban truck trips (Holguín-Veras & Thorson, 2003; Raothanachonkun, Sano, Wisetjindawat, & Matsumoto, 2007). Taken together, these findings suggest that the chaining of individual pickups and deliveries into tours is a phenomenon that cannot be ignored in urban freight modelling. Several models that incorporate such dynamics are discussed in later sections.

8.2.3 Urban Input–Output Data and Models

One challenge is in understanding how economic flows give rise to commodity flows, which in turn, are manifested as vehicular flows. Economic input–output (IO) accounts depict the trading relationships between firms, to include intermediate products, imports and exports (also see Chapter 2). The flows are measured in annual dollar terms and are typically compiled nationally or by regions within a country. They are not as commonly found at the state or urban level, although have been used at that level either through derivation of accounts specific for an urban area or assumption that state or regional relationships hold true for an urban area within it (Hewings, 1985; Jun, 2005). Integrated land use–transport models often use an extension of IO accounts known as social accounting matrices that also incorporate urban land markets (De la Barra, 1989).

IO data, as well as technical (make and use) coefficients derived from them, have been used to quantify the linkages between firms and commodity flows in several recent models. They can be considered a 'push' model of urban freight, in that the flows are depicted as flowing from producing to consuming sectors. They mostly follow a framework originally pioneered by Isard (1951) and Polenske (1974, 1975), where commodities were mapped to the economic sectors that produced and consumed them. Changes in levels of activity within each sector result in a ripple effect in commodity flows. This approach has long been used in statewide or regional commodity flow models (Donnelly, Costinett, & Upton, 1999; Memmott, 1983; Sorratini, 2000), where changes in employment by sector are used as proxies for changes in economic activities.

The use of IO data in urban freight modelling has been rare, in part because trucks have traditionally been modelled rather than the specific commodities carried by them. Significant data limitations and methodological issues also hindered their use in earlier years (Wegener, 2004). Morrison & Smith (1974) extended a national IO model for 20 sectors of the UK economy to the city of Peterborough using a variety of approaches. They eventually settled upon an iterative proportional fitting technique, using estimates of employment in each sector to isolate the IO flows attributable to the target city. Technical coefficients were then derived that were judged to be markedly superior to those produced by other methods. If one accepts that there is little difference between the local and national technical coefficients, the technique offers a cost-effective way to synthesise IO data the urban level. Harris & Liu (1998) found that constraining an urban IO model with data on imports and exports resulted in a measurable improvement in accuracy, with obvious implications for freight modelling.

Jun (2004, 2005) later developed a metropolitan input–output (MIO) model for Seoul that incorporated many elements of social accounting matrices. Starting with an inter-regional IO model, technical coefficients by production location, consumption location and place of residence stratified by income were derived. These coefficients were used to inform destination choice models and other travel choice models, and were used to quantify impacts of various scenarios tested with the model.

Anas & Liu (2007) used an IO model within the larger framework of a dynamic general equilibrium model of Chicago. In contrast to the work cited above, the IO framework is embedded within a larger land use modelling system used to define inter-industry demands. They derived both physical and monetary technical coefficients, which mapped goods and labour market interchanges, respectively. Unlike traditional IO models, the levels of overall demand were endogenous to their model (via Cobb–Douglas production functions) instead of being exogenously defined.

None of these models were developed to specifically study urban freight. However, the possibilities are readily apparent, for they provide a much richer vector of firm interactions than simply the distance between them. By linking sources of production and attraction applicable to each commodity, as revealed in economic surveys, such models illuminate the inter-industry relationships inherent in supply chains. The resulting models should be significantly better than those that rely upon replicating observed trip length distributions alone. Albino, Izoo, & Kuhtz (2002) described an idealised IO model of supply chains and illustrated the concepts through a prototype in Italy. Donnelly (2009) later formulated a stochastic destination choice model that incorporated both trip length and IO-based inter-industry linkages. The resulting model replicated observed O—D patterns much more closely than formulations based solely on the former. Whether used to describe the level of transactions (i.e. annual dollar flows between economic sectors) or the probabilities of inter-industry linkages (i.e. using technical coefficients to identify likely trading partners), such models will play an increasingly prominent role in urban freight modelling.

8.2.4 Firmographic Models and Business Metrics

Substantially more detailed data on firms and employment are needed to effectively use IO data than those used with traditional urban truck models. The latter were almost entirely based on a small number of employment categories. Obtaining the more detailed data on firms, to include industrial classification code, number of employees, size and other attributes will be required in order to use more sophisticated freight modelling techniques. The resources required to develop these data are as large as the models they feed, although advances in geographic information systems and rich spatial data have increased the amount of data about firms and households available for modelling. Most businesses strive to be found in online searches and maps. These data are increasingly being mined to build the detailed databases required for many marketing and advertising purposes in addition to uses in planning and public policy analyses.

A method for synthesising a population of firms was developed by Chiang & Roberts (1976). They described how aggregate data on firm characteristics derived from the US County Business Patterns and IO relationships could be used to generate synthetic firms with size, industry and commodity consumption attributes. They used iterative proportion fitting to estimate the number of firms in each sector by ranges of firm sizes. A means for allocating firms to specific locations below the county level was not discussed, although rule-based methods for doing so, possibly in conjunction with parcel-level land coverage data, are easily envisioned.

De Jong & Ben-Akiva (2007) describe a more elegant approach to firm synthesis that incorporates allocation of aggregate flows to them. Zone-to-zone commodity flows are proportionately allocated to three classes of firms (retailers, wholesalers and manufacturers) based on size characteristics. Their approach goes further than simply allocating the estimates of total production and consumption, however. They also match each consuming firm with one or more producers and assign the commodity flows to them. A variety of probabilistic or sampling approaches can be used to carry out the allocation process. Iterative proportional fitting is used to constrain the allocated flows to balance the production and consumption of flows at each establishment.

Firmographic models have the potential to add considerably more useful details about firms. Such models forecast the birth, death and evolution of every firm within the study area. Moeckel (2006) developed a highly innovative microsimulation approach using Markov models of individual firm transitions. Such methods could be used to simulate the changes in businesses and their location over time. However, Moeckel and others (De Bok & Bliemer, 2006) note that further work is required before such techniques mature, and questions remain as to whether a microsimulation approach is superior to aggregate models. However, their inclusion − or models similar to them − in land use−transport models suggests that they will become more widespread and accessible to freight modellers.

8.2.5   The Supply Chain Context of Urban Freight

Fundamental changes in the way that goods are produced and distributed have taken place over the past three decades. Transport decisions are increasingly being made within the larger supply chain context (Danielis, Rotaris, & Marcucci, 2010; Hesse, 2008), which has a major impact on how goods are moved (pushed) through successive distribution channels. This has revolutionised how businesses operate and conduct business-to-business (B2B) and business-to-consumer (B2C) transactions. A key characteristic about supply chains is that commodities and other inputs to production continually change during the transformation from raw materials and inputs to final consumed products, as shown in Figure 8.3. Many aspects of the commodities often change, from their physical form to value, packaging and transport requirements. The nodes in supply chains are often far apart with many being global in structure.

The modelling of supply chains is a separate field that focuses upon the production and distribution processes involved, which are unique to each chain (Shapiro, 2007). They require detailed data on production processes and costs that are typically available only to the firm(s) involved and are thought to be difficult to generalise (Beamon, 1998; Huang, Lau, & Mak, 2003). Indeed, continual improvements in the underlying business processes can quickly change its dynamics and costs (Agarwal, Shankar, & Tiwari, 2006; Min & Zhou, 2002). In most supply chain models, the details of the transportation function are of secondary concern or represented in a highly abstract manner.

Although not modelled explicitly for each firm, there are several aspects of supply chains that can be usefully represented in urban freight models. The first is the relationship between firms. Many modern freight models match producers with the appropriate consumers of their products using IO data and models, as discussed earlier.

The second important aspect revolves around how costs are represented. Supply chain managers also consider non-transport costs, to include production, inventory, opportunity, and other costs and factors outside the scope of traditional freight models. A firm may choose what appears to be a suboptimal transport choice when considered in isolation, but efficient when viewed within the larger context of total logistics cost. Moreover, such choices are dynamic, as continual changes in technology, productivity, markets and manufacturing result in continually changing supply chains. Some innovation loops are very quick, while others change slowly enough to remain relevant over the long time horizons covered in transportation forecasts.

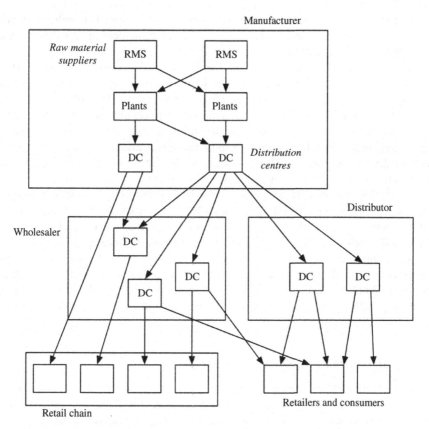

Figure 8.3 Conceptual supply chain actors and interactions.
Source: Adapted from Shapiro, 2007.

Several innovative models base choices on total logistics cost rather than just transport elements. Tavasszy, Smeenk, & Ruijgrok (1998) describes the structure of the SMILE model, which includes production, inventory and transportation costs. The model makes a simultaneous choice of mode and route that minimises the total logistics cost. The formulation is simple yet elegant and has been applied in the Netherlands. The ADA (aggregate–disaggregate–aggregate) model proposed by Ben-Akiva & de Jong (2008) also bases the choice of transport chain (number of stops, mode and vehicle type for each leg, as well as terminals used) on lowest total logistics cost. Their model was calibrated to reproduce observed flows to include empty backhauls. A key drawback to this approach is the reported difficulty in obtaining the data required for model development and application. The model has been applied in Norway, Sweden, Denmark and Flanders.

The third important aspect of supply chains is their extensive use of distribution centres, either as fixed facilities or 'rolling inventories'. Many industries have very lean inventories, limited to the materials used in current production. Much of this material is stored in transit between nodes in the supply chain. The ability to represent such dynamics is increasingly important in freight modelling (Boerkamps, van Binsbergen, & Bovy, 2000; Nagurney, Ke, & Cruz, 2002).

Tardif (2007) found in intercity surveys in Ontario in 1999–2000 that half of the intercity truck tours involved a stop at one distribution centre and another quarter stopped at two (i.e. flows between distribution centres). In further analyses of these data, Donnelly (2009) found that distribution centres in large urban areas were more often the destination than origin, suggesting their emerging role as depots supplying several customers within the urban area. In a sense they have become the next generation of warehouses, with higher frequency of trips to and from them and more frequent private ownership (Baker & Sleeman, 2011).

Distribution centres will play a larger role in B2B and B2C transactions in coming years. Andreoli, Goodchild, & Vitasek (2010) describes the rise of larger distribution centres, capable of serving multiple markets and a wider variety of clients. These will displace older and smaller centres facing technological obsolescence and focus upon single markets. The advent of same-day deliveries from major online retailers will further change the distribution landscape and enlarge B2C channels (Wohlsen, 2013). These will add to the complexity of urban freight by combining military-grade logistics, courier services and retail outlets that double as distribution centres.

8.2.6   Simulation Modelling Frameworks

An ideal framework for modelling urban freight must include the complex dynamics described above in order to provide robust estimates of how changes in demand, policies or infrastructure will affect the transport system. The analytical requirements of models vary considerably from merely understanding the system to testing complex scenarios with second- and third-order economic and environmental effects.

From the standpoint of modelling truck tours, the work of Hunt & Stefan (2007) remains at the forefront. Their model encompassed all commercial trips, of which freight was a subset. The structure of the model is shown in Figure 8.4. Nested logit models (see Chapter 5) are applied at the individual tour level, which are generated as a function of land use rather than economic activity. Tours are not defined or optimised beforehand. Rather, a decision is made at each stop whether to continue on to another destination or return to the origin. The probability of making another stop is calculated in part by the angle formed by the trucks current location, its origin and the location of the next stop chosen from list of all available stops. Stops significantly out-of-direction are rejected in favour of those which move the truck back towards the origin.

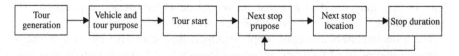

Figure 8.4 Structure of the Calgary commercial vehicle model (Hunt & Stefan, 2007).

A number of models have been proposed that generate tours as a part of the simulation of the larger economic and logistics chain activities they take place within. Both the SMILE model (Tavasszy et al., 1998) in the Netherlands and the ADA modelling approach in Scandinavia (Ben-Akiva & de Jong, 2008), described in earlier chapters, make joint logistic and transport choices within the constraint of total logistics cost. Russo & Carteni (2006) formulated a tour-based urban freight distribution model as a series of nested logit models, proceeding from distribution strategy through first stop choice to subsequent stop choices. As such, it represents a far more holistic approach than the commodity-abstract approach pioneered by Hunt & Stefan (2007). The demand was specified exogenously with the model being successfully applied in Italy.

The current state of the art is in microsimulation of the combined logistics and transport decisions. Monte Carlo simulation approaches are better suited than deterministic models for the high degree of heterogeneity in actors and high degree of variability observed in urban freight. They are highly flexible with respect to modelling, ranging from simple rule-based heuristics to sampling from observed distributions to complex behavioural responses, to include discrete choice models. One of the earliest such models was the GoodTrips model (Boerkamps et al., 2000), which generated goods flows and truck tours as a function of consumer demand. IO relationships were used to link different types of firms in supply chains. Donnelly et al. (1999) developed a hybrid freight model for Oregon that combined an aggregate representation of markets with microsimulation and optimisation of truck tours. Levels of economic activity and IO relationships were exogenously specified using Oregon's integrated land use–transport model. Their model mapped the economic flows into commodity flows by mode of transport, and then allocated them to synthetic firms (including importers and exporters). Discrete shipments were generated and allocated to trucks with multiple stops organised into efficient daily tours. A variant of the model, shown in Figure 8.5, was developed that used estimates of gross urban product by sector for the Portland region in place of the production and consumption estimates provided by the statewide model (Donnelly, 2009).

Wisetjindawat & Sano (2003) described the development of a microsimulation freight model of the Tokyo metropolitan region. A commodity generation model is briefly described, followed by a detailed description of their commodity distribution model. The former is based on linear regression, while the latter is a complex joint model that randomly weights probabilities of distribution channel, location (zone) and shipper choices. The models were estimated using 1982 data from approximately 46,000 firms. They provide evidence, echoed by other researchers, that spatial mixed logit models perform better than other discrete choice formulations. They found that interactions between firms and customers have a larger influence on destination choice than the distance between them.

Samimi, Mohammadian, & Kawamura (2010) are among the many emerging contributors in this area. They have likewise adopted the activity-based modelling paradigm. Their model consists of five modules, as shown in Figure 8.6. They have outlined how publicly available data can be used to populate the model. Like many

Aggregate and disaggregate (deterministic) level

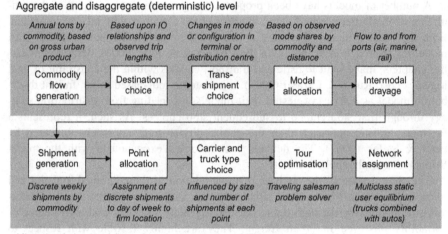

Figure 8.5 Structure of the Mosaic hybrid microsimulation model (Donnelly, 2009).

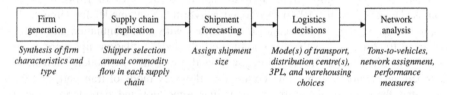

Figure 8.6 Behavioural freight movement microsimulation model (Samimi et al., 2010).

of the frameworks already discussed, logistic decisions inform the transport decisions, which are primarily network based.

This completes our discussion of the push-type models for UFT. As introduced earlier, representing only the actions of producers and flows through supply chains provides an incomplete picture of the dynamics affecting urban freight flows. The demand for goods and services by consumers plays an equally large and significant role. Their consumption patterns and tastes define the markets and commodities that firms compete to supply. They collectively create what we call the 'pull' effect that urban freight models represent and are explored in depth in the following sections.

8.3 Pull Models for Urban Freight

On the basis of the definitions given in the previous sections, we recall some hypotheses that allow description of pull-type behaviour, both in the case of the retailers and the end consumers. Although there are many alternatives, the pull-type behaviours may be summarised as follows:

- the retailer goes to acquisition place, purchases (acquires) the goods, transports the goods to the retail outlet located in the zone d;
- the end consumer arrives at purchasing place, performs the transaction, purchases the commodity, transports the good to the consumption site.

Therefore, the main differences between the two type movements refer to what is the focus of decisional process. In the pull movements, the decision maker, travelling with his/her purchases, combines the choices related to trips and purchases, while in the push-type movements the emphasis is on the transport choices neglecting any relations with purchases.

Note that, at urban scale, the pull models refer to the process of restocking of final products. For this aim, we find different models in the literature that have been developed within the multistep modelling approach (Nuzzolo, Coppola, & Comi, 2013). In the following, some of the main traditional and the new emerging modelling approaches for pull movements of retailer and end consumer are reviewed.

8.3.1 Retailer's Standpoint

The simulation of pull movements performed by the retailer is traditionally modelled using commodity-based modelling through a sequence of models. Following the framework proposed by Russo & Comi (2010), they can be aggregated into two levels (Figure 8.7):

- quantity level concerning estimation of quantity O−D flows of goods; at this level the models entail calculation of
  - attraction flows related to the freight required within each traffic zone of the study area;
  - specialised freight flows (OD matrices) related to the logistics trips (e.g. from the retailer's standpoint);
- vehicle level, which allows quantity flows to be converted into vehicle flows; at this level the models concern determination of

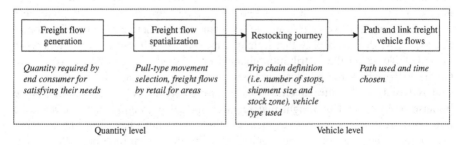

Figure 8.7 Pull-type movements by retailer: modelling structure.

* restocking trip chains in terms of quantity delivered at each stop, zone and vehicle needed for restocking;
* time used as well as the path chosen for restocking sale outlets.

8.3.1.1 The Quantity Level

This level concerns the estimation of freight flows in quantity (e.g. tonnes per day). The quantity flows attracted by each zone are obtained by *attraction* models. The models allow us to obtain the average flow of freight that arrives in each zone of the study area in order to satisfy the freight demand (e.g. shops, food-and-drink outlets). As described in the following section, the attracted freight quantities can also be directly obtained from results of models for shopping trip flows (Gonzalez-Feliu, Toilier, & Routhier, 2010; Oppenheim, 1994; Russo & Comi, 2010). Typically, the attracted quantities are modelled through direct relationships with the attraction power of each zone measured by the number of retailers or sales employees, and land use variables (Kawamura & Miodonski, 2012; Ogden, 1992; Sanchez-Diaz, Holguín-Veras, & Wang, 2013). The most common attraction models are descriptive such as category index models. Assumed to be homogeneous with respect to the freight type, the average quantity of freight attracted (moved) by each retailer or sales employee is directly estimated. The main limitation of category index models is that demand levels are not expressed as function of socio-economic variables other than those used to define categories. Data availability can also restrict the number of categories to be small. Then, the category regression models were proposed. They express the average quantity attracted (moved) by each employee as a function, usually linear, of variables corresponding to the attraction zone.

Once the quantities attracted by each zone have been estimated, it is necessary to spatialise them according to the pull-type movements (*acquisition* model), i.e. the zone, where the retailer brings the freight sold in his/her shop. Therefore, the acquisition model simulates the choice of an origin (or better macroareas) among possible alternatives to bring the freight sold in shops, including commercial concerns (e.g. pubs, restaurants). Typically, the choice of the acquisition macroarea is simulated by random utility models (RUMs, e.g. multinomial logit models: Ibeas, Moura, Nuzzolo, & Comi, 2012; Wisetjindawat, Sano, & Matsumoto, 2005) or gravitational models (Ogden, 1992). The zonal attractiveness variables (such as number of warehouses in the zone and total freight production in the zone) and generalised travel cost are used to explain the attractiveness of the zones. Subsequently, the probability that freight is moved through pull-type movements is modelled. To select the distribution channel or better the hyper-channel (Russo & Comi, 2010) which the retailer purchases, we can assume that the retailer's decision is based on the attractiveness of each freight type (i.e. it affects the choice because bulky freight or freight that requires special packaging may influence its transport), availability of own vehicles (i.e. the vehicle must have some characteristics to allow the transport of the required quantity in a given time period), acquisition place of freight (i.e. it characterises the choice in terms of distance from selling zone). Both descriptive and probabilistic-behavioural models are generally used to simulate this stage. Some empirical results can be found in Danielis et al. (2010), while probabilistic-behavioural models were proposed by Boerkamps et al. (2000), Ibeas et al. (2012), Russo & Comi (2010) and Wisetjindawat et al. (2005), who developed logit models for simulating choices between restocking on their own account or by a third party.

8.3.1.2   The Vehicle Level

The vehicle level allows quantity flows to be converted into vehicle flows. The translation is not direct, particularly in urban areas where freight vehicles undertake complex routing patterns involving trip chains (tours). In fact, each retailer jointly chooses the number and the location of pickups for each tour and hence defines his/her tours, trying to reduce the related costs (e.g. using routing algorithm). Various types of models have been proposed to define restocking tours both based on empirical relationships that allow to combine single trips into a trip chain (Ambrosini, Meimbresse, Routhier, & Sonntag, 2008; Routhier & Toilier, 2007; Sonntag, 1985) and on mathematical relationships. The latter can be further classified according to if the disaggregate or aggregate approach is used. The disaggregate approach involves the use of procedures that include estimation of the restocking tours for each decision maker with different locations of warehouses and shops to restock, different vehicles and time constraints to respect (Crainic, Ricciardi, & Storchi, 2009; Figliozzi, 2006; Gliebe, Cohen, & Hunt, 2007; Polimeni & Vitetta, 2013a; Ruan, Lin, & Kawamura, 2012; Wisetjindawat, Sano, Matsumoto, & Raothanachonkun, 2007). On the other hand, aggregate models consider the average behaviour of all retailers (or categories of retailers) leaving from the same selling zone (Nuzzolo, Crisalli & Comi, 2012).

Within the aggregate approach, the vehicle flows can be obtained using a two-step procedure consisting of simulation of restocking tours through two behavioural models: the *stops per trip chain* and the *size and stock zone* models. Once all restocking tours have been estimated, the freight vehicle O−D flows can be obtained through the aggregation of the trips of the tours.

The stops per trip chain model estimates the number of stops made per trip. In particular, it is possible to simulate how many warehouses are reached for each restocking trip. In the sphere of random utility theory, the trip chain order model can be specified as a logit (Ruan et al., 2012; Smith, Chen, Sana, & Outwater, 2013). The systematic utility function is expressed as a linear function of attributes related to the origin zone (e.g. active accessibility) and to the freight (e.g. type and quantity restocked).

The second model is a joint model that calculates the size and zone of each stop. The shipment size is largely dependent upon the freight type being transported/delivered and the size of recipients (e.g. shop, supermarket and store). The literature contains some models developed for this purpose, but they concern mainly intercity transport (de Jong & Ben-Akiva, 2007; Holguín-Veras, Xu, De Jong, & Maurer, 2011; Rich, Holmbland, & Hansen, 2009). The stock location choice model allows us to define the sequence of zones (i.e. warehouses location) visited during the journey. Logit models are generally used for this choice simulation. The systematic utility includes two groups of attributes: the first considers all variables associated with an alternative destination, such as retailer or warehouse accessibility; the second includes the memory variables representing the history of a tour, such as the cumulative distance covered up to the current location (Kim, Park, & Kim, 2013; Mei, 2013; Nuzzolo & Comi, 2013; Wang & Holguín-Veras, 2008).

The vehicle-type choice has to be investigated as well. Wang & Hu (2012) using RUMs simulated the choice among five vehicle types both for round trip and trip chain. Cavalcante & Roorda (2010) proposed a disaggregate shipment size/vehicle-type choice model based on data collected in the city of Toronto. Nuzzolo & Comi (2013) analysed the joint choice of number of stops per trip chain and of the type of vehicle using data from a survey carried out in the inner area of Rome.

Finally, time and path models should be investigated. In many cities around the world and as confirmed by the literature (Quak & de Koster, 2008; Sathaye, Harley, & Madanat, 2010), time is constrained by governance regulations: the public authorities define one or two time windows (e.g. one in the morning between 8:00 and 10:00 a.m. and one in the afternoon) for the delivery of the goods. For this reason, generally the tour delivery time period model is statistic-descriptive. Furthermore, some authors proposed to take into account the disutility that exists (in general it can assume high values) when there are constraints on departure/arrival time from/to selling zone and to/from warehouse zone. For example, the desired arrival time is not defined as one moment in time, but a time window can be defined in which vehicles can arrive at markets without suffering any penalty. If a vehicle arrives before, it must wait until its time window begins and so it pays a cost. If the vehicle is late, it must pay a penalty proportional to time delay. Some studies on this type of penalty have been developed for urban goods transport. Ando & Taniguchi (2005) show that by multiplying the penalty function and the probability of arrival time, the total early/late penalty can be estimated by a probabilistic model. As regards the path model, in the literature we find many studies developed for ring (one-to-one) trips. Such models are used both for congested networks within equilibrium or dynamic models and for non-congested networks within static or pseudodynamic network loading models. Russo & Vitetta (2003) proposed a specification with a Dial algorithm structure for the implicit assignment of network flows. Further, some models were developed within an optimisation approach. Polimeni & Vitetta (2013b) proposed some vehicle routing models with static travel costs, pseudodynamic travel costs, dynamic travel costs and verified how the cost variation over time influences the route optimisation. Iannò, Polimeni, & Vitetta (2013) proposed an integrated approach for road, transit design in a city logistic plan. The vehicle routing problem for city logistic distribution is designed using optimised road network and the reserved lanes for bus. The whole problem is formulated as a continuous—discrete problem, and the topology and the link capacity are considered. Miao, Qiang, & Ruan (2013) proposed a vehicle routing problem that extends the classical problem by pointing out additional handling costs due to loading and unloading operations for cargo rearrangement with pickups, deliveries and handling costs.

8.3.2   Final Business Standpoint

This class of decision makers includes activities that provide customers with lodging and/or prepared meals, snacks and beverages for immediate consumption (i.e. ho.re.ca.) and manufacturing. The research carried out in this field is very limited, but these types of activities generally attract many deliveries per establishment. Ho.re.ca. represents the majority of activities (especially in the touristic and CBD areas) in many worldwide city centres, and the freight flows destined to final business consumers, in some urban areas, averagely represent the 31% of total daily quantities moved in urban areas (Ibeas et al., 2012; Schoemaker, Allen, Huschebek, & Monigl, 2006). A specific survey carried out in the city of New York (Sanchez-Diaz et al., 2013) revealed that final businesses can determine a high number of movements per day. The daily average number of deliveries per establishment was estimated equal to 2.74, and although they have the lowest level of average sales, they have one of the highest employment levels. At the other hand, this segment of demand is very important because, for example, an establishment located in a high value location will prefer to have more frequent deliveries with fresh food than using a larger area for storage of supplies. Additionally, establishments in this sector are usually small so they have to cope with space constraints to receive goods.

Therefore, even if the ho.re.ca. activities are public shops and final business consumers are end consumers because both of them represent the final destinations for the use of goods, we can assimilate the decision makers to retailers. In fact, due to large quantity of freight daily moved, to the distribution channel used and the place of acquisition, their decisional process is similar to retailer's ones.

As detailed by Danielis et al. (2010) and Stathopoulos, Valeri, & Marcucci (2012) for some Italian cities, the segment of hotels, restaurants and catering (ho.re.ca.) is generally described as a homogenous retail segment. However, the commercial activities present very different logistics and organisational constraints according to the specific service offered. In particular, the share of activities that uses pull-type movements (the so-called cash and carry) is about three/four times higher than traditional shops. Moreover, the used type of movements is strictly related to product type. Own account is relevant for beverage and, in particular, for products whose replenishment can be planned in advance and that can be easy stored (Delle Site, Filippi, & Nuzzolo, 2013). The probability to have pull-type movements decreases when frequency, timing and freight vehicle characteristics are critical issues characterising the logistic activities, while pull-type one prevails if price and variety is particularly relevant.

Then, according to the previous identified behaviour types (i.e. pull- and push-type behaviours) and even if we could be pushed to consider ho.re.ca. and final business consumers belonging to partial push-type class, for the above reasons, their restocking decisional process is closer to retailer than end consumer. Therefore, starting from these statements, Russo & Comi (2013) propose to consider them as retailer and to simulate the choice of the distribution channel and the acquisition place through multinomial and nested logit models. They pointed out ho.re.ca. activities to bring goods for satisfying their customers and on final business consumers to bring goods for satisfying their needs. They found that some movements are freight based (e.g. for chemical products, flowers, hardware, household and hygiene products, the acquisition is performed directly by receivers), while others are strictly related to level-of-service attributes (e.g. travel time, restocking frequency).

8.3.3   End-Consumers' Standpoint

Shopping may be considered a major trip purpose as it forms part of the lifestyle of the population. Shopping is the second most frequent type of urban travel. Nevertheless, most of the current transport literature focuses on studying the characteristics of worker trips, with little emphasis being placed on studying nonworker travel patterns, such as shopping trips (Cao, Chen, & Choo, 2013; Mokhtarian, 2004; Mokhtarian, Ory, & Cao, 2007).

In general, the focus of transportation research is mainly on trip generation, distribution and mode steps within the well-known four-step models (Cascetta, 2009). Although researchers have increasingly emphasised the high incidence of multistop trips in empirically observed behaviour (Dellaert, Arentze, Bierlaire, Borgers, & Timmermans, 1998; Ingene & Ghosh, 1990; Popkowski Leszczyc, Sinha, & Sahgal, 2004; Thill, 1992), the commonly used modelling structure is that described below. In fact, the tour modelling would be less restrictive but has the practical disadvantage of introducing more complexity in data collection and estimation stages. Furthermore, trip chains represent only a limited part of shopping trips, at least in some countries (Arentze & Timmermans, 2001). Note that the four-step modelling was derived especially for simulating commuters' behaviour, and although the commuter behaviour is quite well simulated by this structure, some modifications can be necessary when they are implemented to shopping. For example, generally for commuter, mode choice depends on destination and frequency choices. Upper-level choices (i.e. destination) are actually made taking into account the alternatives available at lower level, such as the modes and the routes to reach each destination. At the other hand, for shopping, the choice of destination where to purchase cannot be related to the mode availability, but only due to characteristics of purchases. For example, the choice to reach a shopping centre, generally located in the suburbs, is mainly related to availability of product, price and brands, and not to the availability of transport mode because generally car remains the main option.

The growth of internet shopping is causing modifications in shopping behaviour. Few studies have examined the geographic distribution of online buyers and its implications on retail development and transport. Cao et al. (2013) found that the influence of shopping accessibility on e-shopping is not uniform, but depends on the locations in metropolitan areas. Specifically, internet users living in urban and/ or high shopping accessibility areas tend to purchase online more often than their counterparts in other areas because the former use the internet more heavily than the latter. However, low shopping accessibility in non-urban areas promotes the usage of e-shopping, compared to non-urban areas with relatively high shopping accessibility.

8.3.3.1   Trip-Based Modelling

The traditional travel demand models simulate the trips making up a journey assuming that the decisions (choices) for each trip are independent of those for other possible trips belonging to the same journey. These assumptions are reasonable when the journey is round trip with a single destination and two symmetric trips. Although, in the recent years, there has been an increasing complexity of the structure of human activities, and therefore travel, especially in urban areas, 70% of the shopping trips are included in home–shopping–home chains (Gonzalez-Feliu, Routhier, & Raux, 2010). Then, the global demand function can be decomposed into submodels, each of which relates to one or more choice dimensions. The sequence most often used is as follows (Figure 8.8): trip generation, shop type and location, mode and path choice.

The trip generation or trip frequency model estimates the mean number of 'relevant' trips undertaken for shopping by the generic end consumer. Trip generation is mainly affected by socio-economic characteristics and land use patterns (or the physical characteristics of the area; Cubukcu, 2001; Yao, Guan, & Yan, 2008). To describe this model, first we define the mean number of trips undertaken by the individual of a given category (average index), then we apply this value to whole set of users belonging to the category. The average index can be estimated by two main categories of models: behavioural (or more properly, RUMs; Russo & Comi, 2012) and descriptive models (Gonzalez-Feliu et al., 2010; Procher & Vance, 2013; Seo, Ohmori, & Harata, 2013). Behavioural models are mainly RUMs and allow to estimate the probability of undertaking one or more than one trip. The systematic utility function includes variables representing the need or the possibility to carry out activities connected with shopping. These variables may relate either to the family or to the individual. Examples of the former type are income and number of members of the household, while examples of individual's variables may be occupational status, gender and age. Other attributes can be related to the origin zone such as active accessibility with respect to the possible destinations for shopping.

The shop type and location model simulates the choice of a type (i.e. local market, hypermarket) and a destination among possible alternatives to purchase;

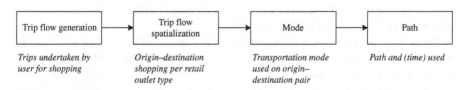

Figure 8.8 Pull-type movements by end consumer: modelling structure.

typically, both probabilities have a multinomial logit structure; there are several methods to model trip distribution, but few of them investigate the choice of shop type (i.e. supermarket, local market; Gonzalez-Benito, 2004). At the other hand, catchment area models can also be used to estimate the main customer's locations (Kubis & Hartman, 2007; Long-Lee & Pace, 2005). Amongst others, Arentze & Timmermans (2001) modelled the choice of the shopping centre where the goods are bought through discrete choice models. Ibrahim (2002) and Jang (2005) used joint disaggregated models to describe the generation and distribution of shopping trips; Gonzalez-Feliu et al. (2012), assuming that the choice of purchasing location occurs simultaneously with modal choice, propose to obtain O−D shopping trips by first using regressive models to simulate the attracted trips for purchasing; second, the origin of trips is simulated by a gravity model; they refer only to the car mode. Veenstra, Thomas, & Tutert (2010) proposed an aggregated method to model trip distribution for the shopping purpose; the model is based on the gravity method, but it takes the spatial configuration of supermarkets into account. Comi & Nuzzolo (2013) presented some logit models that allow to simulate the choice among three different retail outlet types according to four freight types. At this stage, models could consider a huge number of elementary alternative destinations, which does not always conform with realty. The end consumer could choose where to make purchases within a pre-defined and well-known choice set according to some specific attributes (e.g. brands, price). Choice set modelling should also be investigated. In addition, the destination chosen for carrying out an activity (i.e. to buy something) is not a traffic zone but one (or more) elementary destinations (such as a shop) within it. The traffic zone d is therefore a compound alternative consisting of the aggregation of elementary alternatives. Therefore, models can be used considering the size function, as proposed many times building from Ben-Akiva & Lerman (1985).

The mode choice models simulate the fraction (probability) of trips made by end consumers using a given transportation mode. Mode choice is a typical example of a travel choice that can be modified for different trips in which performance or level-of-service attributes have considerable influence. Multinomial logit mode choice models are commonly used, and the systematic utility function is expressed as a function of attributes of a possible transport modes in relation to O−D pair (e.g. travel time and costs, number of wholesalers at zone o) and socio-economic attributes of the end consumer (e.g. gender, income, car availability). In the literature, there has been some research on understanding of shoppers' attitudes towards the various transport modes for shopping purposes (Cervero, 1996; Recker & Stevens, 1976; Williams, 1978); some researchers propose to model destination and mode choices jointly (Richards & Ben-Akiva, 1974; Vrtic, Frohlich, & Schussler, 2007) or mode and departure choices (Bhat, 1998).

Then, path used should be investigated. Path choice behaviour and the model representing it depend on the type of service offered by the different transport modes. Several models have been proposed both for private or public transport. For an overview, we refer to Cascetta (2009) and Nuzzolo, Crisalli, & Rosati (2012).

8.3.3.2   Activity-Based Modelling

Although the assumptions of trip-based modelling are often valid, the number of trips connecting several activities (including shopping) in different locations is increasing. For this reason, the literature proposes some demand models simulating the sequence, or the chain, of trips making up each journey. In particular, some of them proposed to simulate carrying out activities and the related journeys (i.e. activity-based model).

In the activity-based modelling, journey is derived from the demand for activity participation (i.e. the sequences or patterns of behaviour, and not individual trips, are the relevant unit of analysis). Besides, the household and other social structures influence travel and activity behaviour, while spatial, temporal, transportation and interpersonal interdependencies constrain both activity and travel behaviour. The scheduling of activities in time and space is also taken into account. Therefore, the travel demand is derived from the need to pursue activities distributed in space and time (Axhausen & Gärling, 1992; Jones, Koppelman, & Orfeuil, 1990).

According to Gan & Recker (2008), the activity-based models can be classified as models developed within the random utility theory (Ben-Akiva & Bowman, 1995) and models developed as activity scheduling problem (Arentze & Timmermans, 2000; Doherty & Axhausen, 1999). With regards to the former models, a state of the practice for shopping travel demand model has been proposed by Limanond, Niemeier, & Mokhtarian (2005). In particular, a shopping demand model was also proposed, developed within the partial share approach. It consists of various stages that simulate household tour frequency, participating party, shopping tour type, mode and destination choices using a tour-based nested-logit model. The model hence has a structure that can point out the interactions among household members, and the land use effects on decisions. In fact, each decision is conditioned by higher level choices and is influenced by the choices at the lower level via the expected utility (i.e. the logsum).

Some of latter models represent the process of activity scheduling decision making through rule-based structure. In particular, many of them assume some fixed planning order for specifying the activity attributes, while Auld & Mohammadian (2012) proposed an activity-based microsimulation model that simulates activity and travel planning and scheduling. In this context, according to Bhat & Koppelman (2003) the development of an activity-based modelling should consist of: developing representations for activity—travel patterns, developing a comprehensive econometric modelling framework, assembling data and estimating the different model components, developing a microsimulator that uses the econometric modelling system for activity—travel prediction and evaluating the impact of policy actions on activity—travel patterns using the developed simulator.

Note that the activity-based approach leads to an increased complexity in modelling development and in data collection. In fact, it requires time-use survey data for analysis and estimation. A time-use survey entails the collection of data regarding all activities (in-home and out-of-home) pursued by individuals over the course of a day (or multiple days).

8.3.4 The Overall Modelling Framework

Today, there is a growing interest to define tools able to support decision makers to understand the structure of the freight urban system and to compute some indicators that compared with target and benchmarking values allow to identify its level of service. Then, there are models proposing to integrate the previous types of models in a general framework, representing commodity flows as generated by the consumption of the commodity, as a component of the generic urban activity undertaken by consumers. The modelling systems classified above could all be considered to belong to this general framework. The main characteristic of this general framework is the representation of the interacting behaviour of commodity consumers and commodity suppliers/shippers/retailers. Russo & Comi (2010) pointed out the retailer's choice for restocking on his/her own account and propose some behavioural models for its simulation, focusing mainly on logistics chains and on the main assumptions that liveability and accessibility of urban areas are influenced by freight traffic resulting from logistical choices in the supply chain, like warehouse location, delivery frequencies, vehicle type and routing. Boerkamps et al. (2000) proposed a general framework implemented in GoodTrip and applied to the city of Groningen. The developed models simulate these choices and their effects, in current and future situations, through some identified empirical relationships. GoodTrip allows the estimation of goods flows (in terms of quantity), urban freight traffic and its impacts, like vehicle mileage, network loads, emissions and, finally, energy use of urban freight distribution.

In Germany, Sonntag (1985) proposed the support system WIVER that allows to simulate vehicle trip O−D for restocking activities. WIVER starts from the estimation of O−D quantity matrices and provides information regarding total mileage, number of trips and tours, daily traffic distribution over time, subdivided into vehicle type and economic sectors (freight types). Furthermore, the relations between origins and destinations in terms of routes or single trips are modelled through some identified empirical findings. In France, the support system named FRETURB (Ambrosini, Gonzalez-Feliu, & Toilier, 2013; Gonzalez-Feliu et al., 2012; Routhier & Toilier, 2007) was proposed. It uses the movement/delivery (pickup and delivery) as simulation unit. The models implemented within FRETURB consist of a sequence of statistic-descriptive models within three modules which interact with each other: a pickup and delivery model including flows between all the economic activities of a town; a town management module, consisting of transport of goods and raw material for public and construction works, maintenance of urban networks (sewers, water, phone) and garbage; a purchasing trips model, modelling shopping trips by car, which represents the main last kilometre trips to end consumers. The pickup and delivery (generation and attraction) model is a regression-based model. The model has been implemented in about 20 French towns (including Paris, Lyon and Lille).

Based on the same approach implemented in FRETURB and within the European Project CityPorts (City Ports, 2005), CityGoods was developed by Gentile & Vigo (2006). This support system was tested on several cities of the

Emilia-Romagna Region (Italy). The objective was to build a demand generation model in order to estimate the yearly number of operations generated by each zone in terms of tours. On the basis of surveys among transporters, shippers, establishments, they proposed a specific approach for the generation of total number of goods movements as a function of the NACE (European Classification of Economic Activities) code and the number of employees at each establishment. The generation model uses a hierarchical classification of activities of the establishments in a zone. The distribution and network assignment models are in progress.

Finally, Comi & Rosati (2013) proposed a City Logistics Analysis and Simulation support System (CLASS) for the identification of critical stages and the simulation of city logistics scenarios. The analysis of the current scenario focuses on logistics and freight transport in relation to land use, freight restocking demand and supply, logistic profile and road network performances and impacts. The simulation is able to point out the relations existing among city logistics measures, decision maker choice dimensions by using a multistage demand model and a discrete choice approach for each decision level. Among the outputs, the freight vehicle flows on the road network links allow to compute the link performances in terms of congestion, pollution and road accidents involving freight vehicles.

The desire to move activity—travel models into operational practice has roused the interest in microsimulation, a process through which the choices of an individual are simulated stochastically (and sometimes dynamically) based on the underlying models. Partially and fully operational activity-based microsimulation systems include (Bhat, Guo, Srinivasan, & Sivakumar, 2004): Microanalytic Integrated Demographic Accounting System (MIDAS), the Activity Mobility Simulator (AMOS), Prism Constrained Activity—Travel Simulator (PCATS), SIMAP, ALBATROSS, TASHA, Florida's Activity Mobility Simulator (FAMOS) and Comprehensive Econometric Microsimulator for Daily Activity—Travel Patterns (CEMDAP).

8.4   Emerging Modelling Approaches

Agent-based modelling (ABM) is perhaps the ideal platform for implementing urban freight models in a framework that can accommodate the representation of both push and pull dynamics. Cast as an alternative to classical mathematical and statistical approaches, such models are characterised by the autonomy and independence of individual agents. Their behaviours are not constrained by their society, nor are the resulting emergent outcomes predictable in advance. Indeed, Bonabeau (2002) asserts that ABM is a mindset more than a technology, while Epstein (2011) labels it as generative social science. Bankes (2002) notes that ABM allows the relaxation of several unrealistic assumptions often dictated by traditional approaches, such as linearity, homogeneity, normality and stationarity. Such departures seem to well characterise urban freight and the actors and processes influencing it. ABM is a particularly promising approach for representing the interactions of the wide variety of actors in urban freight — shippers, carriers, intermediaries, consumers and third-party logistics firms – all of whom often have conflicting goals, differing information and visibility of the supply chain, and choices open to them.

A precise definition of agents, or of widely accepted methods of modelling them, remains elusive (d'Inverno & Luck, 2003). Most are implemented within ABM modelling platforms that are extensions of the C++ or Java object-oriented programming languages. Autonomous agents are defined as separate threads or processes and situated in a distributed computing environment. Each agent possesses goals, sensors, a means of communicating with adjacent agents, and behavioural constructs that enable them to act upon their perception of their internal state and the environment.

In practice each firm or consumer in an urban area might be considered an agent. Such a 'pure' ABM implementation is not tractable for modelling urban freight, as the sheer number of agents required is computationally intractable. As a consequence, most researchers have either adopted agent concepts using an object-oriented approach or have adopted a hybrid approach. The latter includes cases where objects and agents are mixed, or where only a fraction of the agents are modelled while the remainder are treated as static entities. The MATSim framework (Balmer, Meister, Rieser, Nagel, & Axhausen, 2008) is an example of the latter, although applications of it to freight have not been reported. Elegant conceptual work has been completed by Anand, Yang, van Duin, & Tavasszy (2012) and Roorda, Cavalcante, McCabe, & Kwan (2010) are commended for further reading.

Multilevel models similarly hold considerable promise for extending current models. Xu, Hancock, & Southworth (2003) proposed an elegant multilevel microsimulation model of regional freight flows. The model was novel in that it included three levels of representation, as shown in Figure 8.9. Demand in dollars flow from right to left in the top layer, which results in flows through an idealised supply chain in the middle layer. These, in turn, are translated into commodity flows on a transport layer. Data, to include price and commodity signals, is shared between layers through an information network. A number of simulation models are used to model the activities on each level of the model.

A similar approach has been proposed for use in Chicago, where macroeconomic activities at a national level will place the city within context of its domestic and international trade on the top level of the modelling system. A mesoscopic model incorporating supply chain and transport decisions at a finer level of detail will be used to generate flows within the Chicago region. A microscopic model will be used to assess the network and secondary impacts of the freight flows. Success with a prototype of the mesoscopic model has been obtained (Cambridge Systematics, 2011). Multilevel approaches are well suited for using data and implementing models at levels of spatial and temporal resolution appropriate for each (e.g. trade models, the urban economy, firms and vehicles), rather than forcing them all to operate at a single scale.

Taken together, the current and emerging modelling approaches described in this chapter are capable of meeting a wide range of analytical needs. A number of

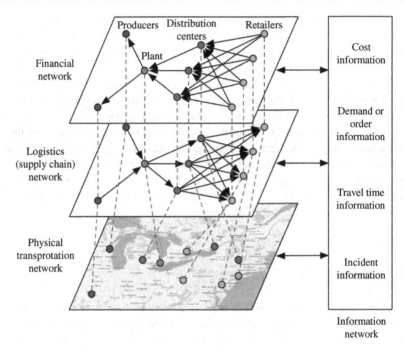

Figure 8.9 A multilevel approach to freight modelling.
Source: Adapted from Xu et al., 2003.

similar projects are under way, as well as equally strong parallel work in the modelling of supply chains. While a single ideal modelling framework is unlikely to emerge the choice can be made between several compelling ones.

In modelling urban freight, some aspects should be further pointed out. The increasing availability of data from cities indicates that the characteristics of urban goods movements mainly depend on the size of urban area and on the level of revealed impacts. For example, as city size and impacts increase, so does the sensitiveness of administrators towards these traffic impacts. In many of these cases, measures that provide to shift toward push-type movements are preferred (e.g. two-tier systems or incentives to switch to third party). Besides, the commercial supply is changing with the advent of new market segments (e.g. low cost products, products at km 0, bioproducts) that push to pull-type movements mainly because price and variety are particularly relevant. Besides, the predominance of such type movements is also strictly related to end-consumer characteristics (e.g. income, age, education) and freight types. For example, the high fashion shops generally have a centralised logistic coordination supplying all the stores included in the network and are based on centralised distribution centres. The product volume is small or medium, while the consignment size is large for seasonal orders concentrated during some yearly periods, while it is small for stock replenishment. Then, retailers are restocked mainly through push-type movements, while end consumers have pull behaviours.

Another aspects refers to the modelling of decision makers' behaviour. Different approaches other than consolidated approaches exemplified by RUMs should be investigated (also see Chapter 5). Of the non-RUM approaches, of interest are those which consider possibility instead of probability, derived from fuzzy-type structures. For example, a comparison between random and fuzzy models for road route choice for national freight transport was given by Quattrone & Vitetta (2011).

Further aspect refers to the growing importance of understanding and modelling the interactions between spatial economic and transport systems. Although the first unifying modelling frameworks to simulate the spatial economic transport interaction process at urban and national scales have been proposed, the future developments should be addressed to open new possibilities for the development of advanced decision support systems focusing on models of localisation of urban distribution centres and large shopping centres.

8.5   Conclusions

This paper presented an overview of models developed for simulating freight transport at the urban scale. The actors and their relevant actions, insofar as they influence urban freight movements, were investigated. Two behaviour types were identified and hence two types of models were reviewed according to these classes of urban goods movements: push and pull models.

The modelling review showed that although significant modelling improvements have been made in recent years, extensions are still needed to deal chiefly with three main issues: data collection, understanding of the decision-making process and integration of pull and push models. Despite the large number of studies on the simulation of UFT, it is widely agreed that the proposed approaches have not yet been fully validated, mainly due to the lack of data. As regards the decision-making process, further research is needed on the interactions among different decision makers involved in urban freight transport and logistics taking into account the constraints and external influences (e.g. market structure, government policy, city structure, traffic management). Modelling frameworks exist both for push and pull movements, but limited research has been done on their integration. In order to move towards model integration, the interactions among decision makers (i.e. agents or actors) should be studied in depth, including interactions between firms (i.e. producers, transport and logistics operators, wholesalers, retailers) of different scales and power levels, firms with different motivations or business objectives, firms with different roles in the system (e.g. shippers, carriers and receivers) and firms versus end consumers. Therefore, research should also focus on developing a general modelling framework, taking into account the mechanism of the location of freight centres/platforms and shopping centres. The above goals could include the study of the dynamic evolution of interactions for both short- and long-term effects: the short-term effects consider the purchase behaviour of end consumers. The long-term effects have to point out land use and transport interactions through, e.g. LUTI-type modelling, focusing on models of location of urban distribution centres and large shopping centres.

Acknowledgements

The authors wish to thank Lori Tavasszy and Gerard de Jong for their inputs and for their useful comments and suggestions on earlier versions of this paper.

References

Agarwal, A., Shankar, R., & Tiwari, M. K. (2006). Modelling the metrics of lean, agile, and leagile supply chain: An ANP-based approach. *Production, Manufacturing and Logistics, 173*(1), 211–225.

Albino, V., Izoo, C., & Kuhtz, S. (2002). Input–output models for the analysis of a local/global supply chain. *International Journal of Production Economics, 78*(2), 119–131.

Ambrosini, C, & Routhier, J. L. (2004). Objectives, methods and results of surveys carried out in the field of urban freight transport: An international comparison. *Transport Reviews, 24*(1), 57–77.

Ambrosini, C., Gonzalez-Feliu, J., & Toilier, F. (2013). A design methodology for scenario-analysis in urban freight modeling. *European Transport/Trasporti Europei, 54*(7).

Ambrosini, C., Meimbresse, B., Routhier, J., & Sonntag, H. (2008). Urban freight policy-oriented modelling in Europe. In E. Taniguchi, & R. G. Thompson (Eds.), *Innovations in city logistics* (pp. 197–212). Hauppauge, NY: Nova Science Publishers.

Anand, N., Quak, H., Van Duin, R., & Tavasszy, L. (2012). City logistics modelling efforts: Trends and gaps–A review. *Procedia–Social and Behavioural Sciences, 39*, 101–115.

Anand, N., Yang, M., van Duin, J. H. R., & Tavasszy, L. (2012). GenCLOn: An ontology for city logistics. *Expert Systems with Applications, 39*(15), 11944–11960.

Anas, A., & Liu, Y. (2007). A regional economy, land use, and transportation model (RELU-TRAN): Formulation, algorithm design, and testing. *Journal of Regional Science, 47*(3), 415–455.

Ando, N., & Taniguchi, E. (2006). An experimentation study on the performance of probabilistic vehicle routing and scheduling with ITS. In E. Taniguchi, & R. G. Thompson (Eds.), *Recent advances in city logistics* (p. 5974). The Netherlands: Elsevier.

Andreoli, D., Goodchild, A., & Vitasek, K. (2010). The rise of mega distribution centers and the impact on logistical uncertainty. *Transportation Letters, 2*, 75–88.

Arentze, T. A., & Timmermans, H. J. P. (2000). *ALBATROSS* – A Learning-Based Transportation Oriented Simulation System. the Netherlands: The European Institute of Relating and Services Studies.

Arentze, T. A., & Timmermans, H. J. P. (2001). Deriving performance indicators from models of multipurpose shopping behaviour. *Journal of Retailing and Consumer Services, 8*, 325–334.

Auld, J., & Mohammadian, A. (2012). Activity planning process in the agent-based dynamic activity planning and travel scheduling model. *Transportation Research Part A, 46*, 1386–1403.

Axhausen, K., & Gärling, T. (1992). Activity-based approaches to travel analysis: Conceptual frameworks, models and research problems. *Transport Reviews, 12*, 324–341.

Baker, P., & Sleeman, J. (2011). The impact of economic and supply chain trends on British warehousing. In *Logistics research network 2011 conference*, Cranfield University. Available from <http://dspace.lib.cranfield.ac.uk/1826/7112/1/The_impact_of_economic. pdf> Accessed 02.04.13.

Balmer, M., Meister, K., Rieser, M., Nagel, K., & Axhausen, K.W. (2008), *Agent-based simulation of travel demand: Structure and computational performance of MATSim-T*. Available from <http://matsim.org/uploads/BalmerEtAl2008ITM.pdf> Accessed 18.05.13.

Bankes, S. C. (2002). Agent-based modeling: A revolution? *Proceedings of the National Academy of Science, 99*(3), 7199–7200.

Beamon, B. M. (1998). Supply chain design and analysis: Models and methods. *International Journal of Production Economics, 55*(3), 281–294.

Ben-Akiva, M., & Bowman, J. L. (1995). *Activity based disaggregate travel demand model system with daily activity schedules*. International conference on activity based approaches: Activity scheduling and the analysis of activity patterns, *May 25–28*. the Netherlands: Eindhoven University of Technology.

Ben-Akiva, M., & de Jong, G. (2008). The aggregate–disaggregate–aggregate (ADA) freight model system. In M. Ben-Akiva, H. Meersman, & E. Van De Voorde (Eds.), *Recent developments in transport modelling*. London: Emerald Publishing.

Ben-Akiva, M., & Lerman, S. R. (1985). *Discrete choice analysis: Theory and application to travel demand*. Cambridge, MA, USA: MIT Press.

Bhat, C., Guo, J. Y., Srinivasan, S., & Sivakumar, A. (2004). *A comprensive econometric micro-simulator for daily activity–travel patterns (CEMDAP)*. *Proceedings of the 83rd Transportation Research Board annual meeting*. Washington, DC, USA: Transportation Research Board of the National Academies.

Bhat, C., & Koppelman, F. (2003). Activity-based modeling of travel demand. In R. Hall (Ed.), *Handbook of transportation science (3)* (pp. 39–66). Dordrecht: Kluwer Academic Publishers.

Bhat, C. R. (1998). Analysis of travel mode and departure time choice for urban shopping trips. *Transportation Research Part B, 32*(6), 361–371.

Boerkamps, J. H. K., van Binsbergen, A. J., & Bovy, P. H. L. (2000). Modelling behavioural aspects of urban freight movement in supply chains. *Transportation Research Record, 1725*, 17–25.

Bonabeau, E. (2002). Agent-based modelling: Methods and techniques for simulating human systems. *Proceedings of the National Academy of Sciences, 99*(3), 7280–7287.

Cambridge Systematics, Inc. (2007). *Quick Response Freight Manual II*, Publication No. FHWA-HOP-08-010, Federal Highway Administration, Washington, DC.

Cambridge Systematics, Inc. (2011). *A working demonstration of a mesoscale freight model for the Chicago region*. Chicago: Chicago Metropolitan Agency for Planning.

Cao, X, Chen, Q., & Choo, S. (2013). *Geographical distribution of e-shopping: An application of structural equations models in the twin cities*. *Proceedings of the 92nd Transportation Research Board annual meeting*. Washington, DC, USA: Transportation Research Board of the National Academies.

Cascetta, E. (2009). *Transportation systems analysis: Models and applications*. Newyork: Springer.

Cavalcante, R., & Roorda, M. J. (2010). *A disaggregate urban shipment size/vehicle-type choice model*. *Proceedings of the 89th Transportation Research Board annual meeting*. Washington, DC, USA: Transportation Research Board of the National Academies.

Cervero, R. (1996). Mixed land-use and commuting: Evidence from the American housing survey. *Transportation Research A*361–377.

Chiang, Y. S. & Roberts, P. O. (1976). Representing industry and population structure for estimating freight flows. *Report 76-8*, Centre for Transportation Studies, Massachusetts Institute of Technology.

City Ports. (2005). *City Ports — Project interim report*. Emilia-Romagna Region, Bologna, Italy.

Comi, A., & Nuzzolo, A. (2013). *Simulating urban freight flows with combined shopping and restocking demand models. Proceedings of the 8th conference on city logistics.* Kyoto, Japan: Institute for City Logistics.

Comi, A., & Rosati, L. (2013). CLASS: A city logistics analysis and simulation support system. *Procedia — Social and Behavioral Sciences.*

Crainic, T. G., Ricciardi, N., & Storchi, G. (2009). Models for evaluating and planning city logistics systems. *Transportation Science, 43*(4), 432−454.

Cubukcu, K. M. (2001). Factors affecting shopping trip generation rates in metropolitan areas. *Studies in Regional and Urban Planning, 9*, 51−68.

d'Inverno, M., & Luck, M. (2003). *Understanding agent systems.* Berlin: Springer-Verlag.

Danielis, R., Rotaris, L., & Marcucci, E. (2010). Urban freight policies and distribution channels: A discussion based on evidence from Italian cities. *European Transport/Trasporti Europei, 46*, 114−146.

Davidson, W., Donnelly, R., Vovsha, P., Freedman, J., Ruegg, S., Hicks, J., et al. (2007). Synthesis of first practices and operational research approaches in activity-based travel demand modelling. *Transportation Research Part A, 41*(5), 464−488. Elsevier.

De Bok, M., & Blierner, M. C. J. (2006). Infrastructure and firm dynamics: Calibration of microsimulation model for firms in the Netherlands. *Transportation Research Record, 1977*, 132−144.

de Jong, G., & Ben-Akiva, M. (2007). A micro-simulation model of shipment size and transport chain choice. *Transportation Research Part B, 41*(9), 950−965.

De la Barra, T. (1989). *Integrated land use and transport modelling: Decision chains and hierarchies.* Cambridge: Cambridge University Press.

Dellaert, B. G. C., Arentze, T. A., Bierlaire, M., Borgers, A. J., & Timmermans, H. J. P. (1998). Investigating consumers' tendency to combine multiple shopping purposes and destinations. *Journal of Marketing Research, 35*, 177−188.

Delle Site, P., Filippi, F., & Nuzzolo, A. (2013). *Linee Guida dei Piani di Logistica Urbana Sostenibile.* Rimini, Italy: Maggioli Editore.

Doherty, S. T., & Axhausen, K. W. (1999). The development of a unified framework for the household activity−travel scheduling process. In W. Brilon, F. Huber, M. Schreckengerg, & H. Wallentowitz (Eds.), *Traffic and mobility: Simulation−economics−environment* (pp. 35−56). Berlin: Springer.

Donnelly, R. (2006). Evaluation of practice today. In K. Hancock (Ed.), *Freight demand modelling: Tools for public-sector decision making, conference proceedings* (40). Washington, DC: Transportation Research Board pp. 27−30.

Donnelly, R. (2009). *A hybrid microsimulation model of urban freight transport demand* (Ph.D. thesis). University of Melbourne.

Donnelly, R., Costinett, P. J., & Upton, W. (1999). *The Oregon statewide and substate travel forecasting models. Statewide travel demand forecasting, Transportation Research Circular E-C011.* Washington, DC: Transportation Research Board.

Epstein, J. (2011). *Generative social science: Studies in agent-based computational modelling.* Princeton: Princeton University Press.

Figliozzi, M. (2006). Modeling the impact of technological changes on urban commercial trips by commercial activity routing type. *Transportation Research Record: Journal of the Transportation Research Board* 118−126.

Figliozzi, M. A. (2007). Analysis of the efficiency of urban commercial vehicle tours: Data collection, methodology, and policy implications. *Transportation Research Part B, 41* (9), 1014–1032.

Figliozzi, M.A., Kingdon, L., & Wilkitzki, A. (2007). *Analysis of freight tours in a congested urban area using disaggregated data: Characteristics and data collection challenges.* Available from <http://www.metrans.org/nuf/2007/documents/Figliozzifreighttours.pdf> Accessed 12.03.10.

Gan, P. L., & Recker, W. (2008). A mathematical programming formulation of the household activity rescheduling problem. *Transportation Research Part B, 42,* 571–606.

Gentile, G., & Vigo, D. (2006). A demand model for freight movements based on a tree classification of the economic activities applied to city logistic citygood. In *Presentation in the 2nd roundtable, BESTUFS workshop TFH,* Wildau.

Gliebe, J., Cohen, O., & Hunt, J. D. (2007). Dynamic choice model of urban commercial activity patterns of vehicles and people. *Transportation Research Record: Journal of the Transportation Research Board, 2003,* 17–26.

Gonzalez-Benito, O. (2004). Random effects choice models: Seeking latent predisposition segments in the context of retail store format selection. *Omega, 32,* 167–177.

Gonzalez-Feliu, J., Ambrosini, C., Pluvinet, P., Toilier, F., & Routhier, J. L. (2012). A simulation framework for evaluating the impacts of urban goods transport in terms of road occupancy. *Journal of Computational Science, 3*(4), 206–215.

Gonzalez-Feliu, J.Routhier, J.L., & Raux, C. (2010). An attractiveness-based model for shopping trips in urban areas. In *Proceedings of 12th world conference on transport research,* Lisbon, Portugal.

Gonzalez-Feliu, J., Toilier, F., & Routhier, J. L. (2010). End-consumer goods movement generation in French medium urban areas. *Procedia – Social and Behavioral Sciences, 2*(3), 6189–6204.

Harris, R. I. D., & Liu, A. (1998). Input–output modelling of the urban and regional economy: The importance of external trade. *Regional Studies, 32*(9), 851–862.

Hesse, M. (2008). *The city as a terminal: The urban context of logistics and freight transport.* Aldershot: Ashgate Publishing Ltd.

Hewings, J. G. D. (1985). *Regional input–output analysis.* Beverly Hills, CA: Sage Publications.

Holguín-Veras, J., & Patil, G. (2005). Observed trip chain behaviour of commercial vehicles. *Transportation Research Record, 1906,* 74–80.

Holguín-Veras, J., & Patil, G. (2008). A multicommodity integrated freight origin–destination synthesis model. *Networks and Spatial Economics, 8*(2–3), 309–326.

Holguín-Veras, J., & Thorson, E. (2003). Modelling commercial vehicle empty trips with a first-order trip chain model. *Transportation Research Part B, 37*(1), 129–148.

Holguín-Veras, J., Xu, N., De Jong, G., & Maurer, H. (2011). An experimental economics investigation of shipper–carrier interactions in the choice of mode and shipment size in freight transport. *Networks and Spatial Economics* 1–24.

Huang, G. Q, Lau, J. S. K., & Mak, K. L. (2003). The impacts of sharing production information on supply chain dynamics: A review of the literature. *International Journal of Production Research, 41*(7), 1483–1517.

Hunt, J. D., & Stefan, K. J. (2007). Tour-based microsimulation of urban commercial movements. *Transportation Research Part B, 41*(9), 981–1013.

Hunt, J. D., Stefan, K. J., & Brownlee, A. T. (2006). Establishment-based survey of urban commercial vehicle movements in Alberta, Canada: Survey design, implementation, and results. *Transportation Research Record, 1957*, 75–83.

Iannò, D., Polimeni, A., & Vitetta, A. (2013). An integrated approach for road, transit design in a city logistic plan: A case study. *WIT Transactions on the Built Environment, 130*, 811–822.

Ibeas, A., Moura, J. L., Nuzzolo, A., & Comi, A. (2012). Urban freight transport demand: Transferability of survey results analysis and models. *Procedia – Social and behavioral science, 54*, 1068–1079. Available from http://dx.doi.org/10.1016/j.sbspro.2012.09.822.

Ibrahim, M. F. (2002). Disaggregating the travel components in shopping centre choice. An agenda for valuation practices. *Journal of Property Investment and Finance, 20(3)*, 277–294.

Ingene, C. A., & Ghosh, A. (1990). Consumer and producer behaviour in a multipurpose shopping environment. *Geographical Analysis, 22(1)*, 70–93.

Isard, W. (1951). Interregional and regional input–output analysis: A model of a space-economy. *The Review of Economics and Statistics, 33(4)*, 318–328.

Jang, T. Y. (2005). Count data models for trip generation. *Journal of Transportation Engineering, 6*, 444–450.

Jones, P. M., Koppelman, F. S., & Orfeuil, J. P. (1990). *Activity analysis: State of the art and future directions* (pp. 34–55). Developments in dynamic and activity-based approaches to travel analysis. Aldershot, England: Gower.

Jun, M. (2004). A metropolitan input–output model: Multisectoral and multispatial relations of production, income formation, and consumption. *Annals of Regional Science, 38(1)*, 131–147.

Jun, M. (2005). Forecasting urban land use demand using a metropolitan input–output model. *Environment and Planning A, 37(7)*, 1311–1328.

Kawamura, K., & Miodonski, D. (2012). Examination of the relationship between built environment characteristics and retail freight delivery. In *Proceedings of the 91st annual meeting of Transportation Research Board (TRB)*, Washington, DC, USA.

Kim, H., Park, D., Kim, C., & Park (2013). *Tour-based truck destination choice behaviour incorporating agglomeration and competition effects in Seoul. Proceedings of the 8th conference on city logistics.* Kyoto, Japan: Institute for City Logistics.

Kubis, A., & Hartman, M. (2007). Analysis of location of large-area shopping centres. A probabilistic gravity model for the Halle-Leipzig area. *Jahrbuch für Regionalwissenschaft, 27*, 43–57.

Kuzmyak, J. R. (2008). *Forecasting metropolitan commercial and freight travel. NCHRP Synthesis 384, National Cooperative Highway Research Program.* Washington, DC: Transportation Research Board.

Limanond, T., Niemeier, D. A., & Mokhtarian, P. L. (2005). Specification of a tour-based neighborhood shopping model. *Transportation, 32*, 105–134.

Lindholm, M. (2013). Urban freight transport from a local authority perspective – a literature review. *European Transport/Trasporti Europei, 54(3)*.

List, G. F., Konieczny, L., Durnford, C. L., & Papayanoulis, V. (2001). Best practice truck flow estimation model for the New York City region. *Transportation Research Record, 1790*, 97–103.

Long-Lee, M., & Pace, K. (2005). Spatial distribution of retail sales. *The Journal of Real Estate Finance and Economics, 31(1)*, 53–69.

Maejima, T. (1979). An application of continuous spatial models to freight movements in greater London. *Transportation, 8*(1), 51−63.

Mei, B. (2013). A destination choice model for commercial vehicle movements in the metropolitan area. In *Proceedings of the 92nd Transportation Research Board annual meeting*, Washington, DC, USA.

Memmott, F. W. (1983). *Application of statewide freight demand forecasting techniques. NCHRP Report 260, National Cooperative Highway Research Program.* Washington, DC: Transportation Research Board.

Miao, L., Qiang, M., & Ruan, Q. (2013). *A vehicle routing problem with pickups, deliveries and handling costs. Proceedings of the 92nd Transportation Research Board annual meeting.* Washington, DC, USA: Transportation Research Board of the National Academies.

Min, H., & Zhou, G. (2002). Supply chain modelling: Past, present and future. *Computers & Industrial Engineering, 43*(1−2), 231−249.

Moeckel, R. (2006). *Business location decisions and urban sprawl: A microsimulation of business relocation and firmography* (Ph.D. thesis). University of Dortmund.

Mokhtarian, P. L. (2004). A conceptual analysis of the transportation impacts of B2C e-commerce. *Transportation, 31*(3), 257−284.

Mokhtarian, P. L., Ory, D. T., & Cao, X. (2007). Shopping-related attitudes: A factor and cluster analysis of northern California shoppers. In *Proceedings of 11th WCTR*, Berkeley, CA, USA.

Morrison, W. I., & Smith, P. (1974). Nonsurvey input−output techniques at the small area level: An evaluation. *Journal of Regional Science, 14*(1), 1−14.

Muñuzuri, J., Larrañeta, J., Onieva, L., & Cortés, P. (2004). Estimation of an origin−destination matrix for urban freight transport: Application to the City of Seville. In E. Taniguchi, & R. Thompson (Eds.), *Logistics systems for sustainable cities* (pp. 67−81). Kyoto: Institute for City Logistics.

Nagurney, A., Ke, K., & Cruz, J. (2002). Dynamics of supply chains: A multilevel (logistical−informational−financial) network perspective. *Environment and Planning B, 29*(6), 795−818.

Nuzzolo, A., Crisalli, U., & Comi, A. (2012). A system of models for the simulation of urban freight restocking tours. *In procedia−social and behavioral sciences, 39*, 664−676. Available from < http://dx.doi.org/10.1016/j.sbspro.2012.03.138 >.

Nuzzolo, A., & Comi, A. (2013). Tactical and operational city logistics: Freight vehicle flow modeling. In M Ben Akiva, E. Van de Voorde, & H. Meersman (Eds.), *Freight Transport Modelling* 21, (pp. 433−451). Emerald Group Publishing Limited.

Nuzzolo, A., Coppola, P., & Comi, A. (2013). Freight transport modeling: Review and future challenges. *International Journal of Transport Economics, XL*(2), 151−181.

Nuzzolo, A., Crisalli, U., & Rosati, L. (2012). A schedule-based assignment model with explicit capacity constraints for congested transit networks. *Transportation Research Part C, 20*(1), 16−33.

Ogden, K. W. (1992). *Urban goods movement.* Hants, England: Ashgate.

Oppenheim, N. (1994). *Urban travel demand modelling.* New York: John Wiley & Sons.

Polenske, K. (1974). Interregional analysis of U.S. commodity freight shipments, *Report SP-389*, Society of Automotive Engineers, New York.

Polenske, K. (1975). Multiregional interactions between energy and transportation. In K. Polenske, & J. Skolka (Eds.), *Advances in input−output analysis* (pp. 433−460). Cambridge: Ballinger Publishing Company.

Polimeni, A., & Vitetta, A. (2013a). Optimising waiting at nodes in time-dependent networks: Cost functions and applications. *Journal of Optimization Theory and Applications, 156(3)*, 805−818.

Polimeni, A., & Vitetta, A. (2013b). A comparison of vehicle routing approaches with link costs variability: An application for a city logistic plan. *WIT Transactions on the Built Environment, 130*, 823−834.

Popkowski Leszczyc, P. T. L., Sinha, A., & Sahgal, A. (2004). The effect of multi-purpose shopping on pricing and location strategy for grocery stores. *Journal of Retailing, 80*, 85−99.

Procher, V., & Vance, C. (2013). *Who does the shopping? German time-use evidence, 1996−2009*. Proceedings of the 92nd Transportation Research Board annual meeting. Washington, DC, USA: Transportation Research Board of the National Academies.

Quak, H. J., & de Koster, M. B. M. (2008). Delivering goods in urban areas: How to deal with Urban policy restrictions and the environment. *Transportation Science, 43*, 211−227.

Quattrone, A., & Vitetta, A. (2011). Random and fuzzy utility models for road route choice. *Transportation Research Part E: Logistics and Transportation Review, 47(6)*, 1126−1139.

Raothanachonkun, P., Sano, K., Wisetjindawat, W., & Matsumoto, S. (2007). Estimating truck trip origin−destination with commodity-based and empty trip models. *Transportation Research Record, 2008*, 43−50.

Recker, W., & Stevens, R. (1976). Attitudinal models of modal choice: The multinomial case for selected nonwork trips. *Transportation, 5*, 355−375.

Rich, J., Holmbland, P. M., & Hansen, C. O. (2009). A weighted logit freight mode-choice model. *Transportation Research Part E, 45*, 1006−1019.

Richards, M., & Ben-Akiva, M. (1974). A simultaneous destination and mode choice model for shopping trips. *Transportation, 3*, 343−356.

Roorda, M., Cavalcante, R., McCabe, S., & Kwan, H. (2010). A conceptual framework for agent-based modelling of logistics services. *Transportation Research Part E, 46(1)*, 18−31.

Routhier, J.L., & Toilier, F., (2007). FRETURBV3, a policy oriented software of modelling urban goods movement. In *Proceedings of the 11th world conference on transport research*, Berkeley, CA.

Ruan, M., Lin, J., & Kawamura, K. (2012). Modelling urban commercial vehicle daily tour chaining. *Transportation Research Part E, 48*, 1169−1184.

Ruiter, E. R. (1992). Phoenix commercial vehicle survey and travel models. *Transportation Research Record, 1364*, 144−151.

Russo, F, & Cartenì, A. (2006). Application of a tour-based model to simulate freight distribution in a large urbanized area. In E. Taniguchi, & R. G. Thompson (Eds.), *Recent advances in city logistics* (pp. 31− 45). The Netherlands: Elsevier Ltd.

Russo, F. (2013). Modelling behavioural aspects of urban freight movement. In E. Van de Voorde, M. Ben-Akiva, & H. Meersman (Eds.), *Freight transport modelling* (pp. 353−375). Bingley, UK: Emerald Group Publishing Limited.

Russo, F., & Comi, A. (2010). A modelling system to simulate goods movements at an urban scale. *Transportation, 37(6)*, 987−1009. Available from http://dx.doi.org/10.1007/s11116-010-9276-y.

Russo, F., & Comi, A. (2011). Measures for sustainable freight transportation at urban scale: Expected goals and tested results in Europe. *Journal of Urban Planning and Development, 137(2)*, 142−152. Available from http://dx.doi.org/10.1061/(ASCE)UP.1943-5444.0000052.

Russo, F., & Comi, A. (2012). The simulation of shopping trips at urban scale: Attraction macro-model. *Procedia — Social and Behavioral Sciences, 39*, 387—399. Available from http://dx.doi.org/10.1016/j.sbspro.2012.03.116.

Russo, F., & Comi, A. (2013). A model for simulating urban goods transport and logistics: The integrated choice of ho.re.ca. activity decision-making and final business consumers. In: *Procedia—Social and Behavioral Sciences, 80*, 717—728, *Proceedings of the 20th international symposium on transportation and traffic theory (ISTTT'20).* Available from http://dx.doi.org/10.1016/j.sbspro.2013.05.038.

Russo, F., & Musolino, G. (2012). A unifying modelling framework to simulate the spatial economic transport interaction process at urban and national scales. *Journal of Transport Geography, 24*, 189—197.

Russo, F., & Vitetta, A. (2003). An assignment model with modified Logit, which obviates enumeration and overlapping problems. *Transportation, 30*, 177—201.

Samimi, A., Mohammadian, A., & Kawamura, K. (2010). A behavioural freight movement microsimulation model: Method and data. *Transportation Letters, 2*, 53—62.

Sanchez-Diaz, I., Holguín-Veras, J., & Wang, C. (2013). *Assessing the role of land-use, network characteristics and spatial effects on freight trip attraction. Proceedings of 92nd TRB annual meeting.* Washington, DC, USA: Transportation Research Board of the National Academies.

Sathaye, N., Harley, R., & Madanat, S. (2010). *Unintended environmental impacts of nighttime freight logistics activities, Transportation Research Part A* (44, pp. 642—659). Elsevier.

Schlappi, M. L., Marshall, R. G., & Itamura, I. T. (1993). Truck travel in the San Francisco Bay area. *Transportation Research Record, 1383*, 85—94.

Schoemaker, J., Allen, J., Huschebek, M., & Monigl, J. (2006). *Quantification of urban freight transport effects. I. BESTUFS consortium.* <www.bestufs.net>.

Seo, S., Ohmori, N., & Harata, N. (2013). *Effects of household structure on elderly grocery shopping behaviour in Korea. Proceedings of the 92nd Transportation Research Board annual meeting.* Washington, DC, USA: Transportation Research Board of the National Academies.

Shapiro, J. F. (2007). *Modelling the supply chain* (2nd ed.). Toronto: Duxbury Press.

Slavin, H. (1979). *The transport of goods and urban spatial structure* (Ph.D. thesis). University of Cambridge.

Smith, C., Chen, J., Sana, B., & Outwater, M. (2013). A disaggregate tour-based truck model with simulation shipment allocation to trucks. In *Proceedings of the 92nd Transportation Research Board annual meeting*, Washington, DC, USA.

Sonntag, H. (1985). A computer model of urban commercial traffic. *Transport, Policy and Decision Making, 3*(2), 171—180.

Sorratini, J. A. (2000). Estimating statewide truck trips using commodity flows and input—output coefficients. *Journal of Transportation and Statistics, 3*(1), 53—67.

Southworth, F. (1982). An urban goods movement model: Framework and some results. *Papers in Regional Science, 50*(1), 165—184.

Spiess, H. (1987). A maximum-likelihood model for estimating origin destination matrices. *Transportation Research Part B, 21*(5), 395—412.

Stathopoulos, A., Valeri, E., & Marcucci, E. (2012). *Stakeholder reactions to urban freight policy innovation, Journal of Transport Geography* 22, (pp. 34—45). . Elsevier.

Tardif, R. (2007). *Using operational truck location data to improve understanding of freight flows* (pp. 11—14). *North American freight transportation data workshop,*

Transportation Research Circular E-C119. Washington, DC: Transportation Research Board.

Tavasszy, L., Smeenk, B., & Ruijgrok, C. (1998). A DSS for modelling logistic chains in freight transport policy analysis. *International Transactions in Operational Research, 5* (6), 447–459.

Taylor, S., & Button, K. (1999). Modelling urban freight: What works, what doesn't work? In E. Taniguchi, & R. Thompson (Eds.), *City logistics I* (pp. 203–217). Kyoto: Institute of Systems Science Research.

Thill, J. C. (1992). Spatial duopolistic competition with multipurpose and multistop shopping. *Annals of Regional Science, 26*, 287–304.

Turnquist, M. (2006). Characteristics of effective freight models. In K. Hancock (Ed.), Freight demand modelling: Tools for public-sector decision making, *conference proceedings* (40, pp. 11–16). Washington, DC: Transportation Research Board.

Veenstra, S. A., Thomas, T., & Tutert, S. I. A. (2010). Trip distribution for limited destinations: A case study for grocery shopping trips in the Netherlands. *Transportation, 37*, 663–676.

Visser, J., Nemoto, J., & Browne, M. (2013). *Home delivery and the impacts on urban freight transport: A review*. Proceedings of the 8th conference on city logistics. Kyoto, Japan: Institute for City Logistics.

Vrtic, M., Frohlich, P., Schussler, N., Axhausen, K. W., Lohse, D., Schiller, C., et al. (2007). Two-dimensionally constrained disaggregate trip generation, distribution and mode choice model: Theory and application for a Swiss national model. *Transportation Research Part A, 41*, 857–873.

Wang, Q., & Holguín-Veras, J. (2008). An investigation on the attributes determining trip chaining behaviour in hybrid micro-simulation urban freight models. In *Proceedings of the 87th Transportation Research Board annual meeting*, Washington, DC, USA.

Wang, Q., & Hu, J. (2012). Behavioural analysis of commercial vehicle mode choice decisions in urban areas. In *Proceedings of the 91st annual meeting of Transportation Research Board (TRB)*, Washington, DC, USA.

Wegener, M. (2004). Overview of land-use transport models. In D. A. Hensher, & K. Button (Eds.), *Transport geography and spatial systems* (pp. 127–146). Kidlington: Elsevier Science.

Wigan, M. (1971). *Benefit assessment for network traffic models and application to road pricing* (417). Report LR. Crowthorne, Berkshire, UK: Road Research Laboratory.

Wigan, M.R., & Southworth, F. (2006). What's wrong with freight models, and what should we do about it? *Transportation Research Board annual meeting paper* 06-1757.

Williams, M. (1978). Factors affecting modal choice decision in urban travel. *Transportation Research, 12*, 91–96.

Wisetjindawat, W., & Sano, K. (2003). A behavioural modelling in micro-simulation for urban freight transportation. *Journal of the Eastern Asia Society for Transportation Studies, 5*, 2193–2208.

Wisetjindawat, W., Sano, K., & Matsumoto, S. (2005). Supply chain simulation for modelling the interactions in freight movement. *Journal of the Eastern Asia Society for Transportation Studies, 6*, 2991–3004.

Wisetjindawat, W., K. Sano, S. Matsumoto and P. Raothanachonkun (2007). Micro-simulation model for modelling freight agents interactions in urban freight movement. In *Proceedings of the 86th Transportation Research Board annual meeting*, Washington, DC, USA.

Wohlsen, M. (2013). How robots and military-grade algorithms make same-day delivery possible. *Wired Magazine.* Available from <http://www.wired.com/business/2013/03/online-retailers-faster-than-overnight/all/> Accessed 2.04.13.

Xu, J., Hancock, K., & Southworth, F. (2003). Simulation of regional freight movement with trade and transportation multinetworks. *Transportation Research Record, 1854,* 152−161.

Yao, L., Guan, H., & Yan, H. (2008). Trip generation model based on destination attractiveness. *Tsinghua Science and Technology, 13(5),* 632−635.

9 Freight Service Valuation and Elasticities

Gerard de Jong

Institute for Transport Studies, University of Leeds, UK; Significance BV, The Hague, The Netherlands; and Centre for Transport Studies, VTI/KTH, Stockholm, Sweden

9.1 Introduction

The main practical reason why so many studies have been carried out to derive freight service valuations (and even many more in passenger transport), such as values of transport time (VTT) and values of transport time variability (VTTV) is that in many countries, plans for transport infrastructure projects (new roads and railways, wider roads, more tracks, bridges, tunnels, locks, etc.) and transport policy measures (e.g. road pricing) are evaluated ex ante using cost-benefit analysis (CBA). This requires a conversion of benefits and disbenefits that by their nature are not in money units into money. Important examples are the transport time benefits of infrastructure projects and transport polices to the freight sector, but there can be other impacts that are related to freight, such as reduced transport time variability, increased flexibility, reductions in harmful emissions from freight transport, etc.

A secondary practical reason to undertake freight service valuation studies is that some components of freight transport models need a conversion factor from outside the model, because it is not possible, inefficient or inconsistent (with regards to other components) to estimate the conversion factor within the model component itself. This usually also refers to a situation where a researcher wants to include both transport time and cost, and some exogenous VTT is imputed to have a common denominator for these. But in principle all the other variables mentioned above could be included here.

The values for these two different purposes are not necessarily the same. For the second purpose, everything should be included that affects the choices of agents in freight transport. In CBA carried out for national, supranational or regional authorities (social CBA), some impacts (e.g. taxes) might be excluded on the grounds that these are just transfers between agents in the system.

In Section 9.2, we will discuss the methods that can be used to derive such values, as well as the type of data that are used for this. Several of the models discussed in earlier chapters (such as mode choice models in Chapter 6) can also be used to derive freight valuations. In this Chapter 9, these possibilities will be briefly mentioned, but for a more detailed discussion of these models we will refer

Modelling Freight Transport. DOI: http://dx.doi.org/10.1016/B978-0-12-410400-6.00009-4

to the relevant chapter. Section 9.2 also contains an overview of outcomes, with special sections on VTT and VTTV.

Elasticities give the impact of a change in an exogenous variable (the 'stimulus') on an endogenous variable (the 'response') where both are measured in percentage changes relative to the initial level. The stimulus can for instance be a change in the transport cost of road transport or in the rail transport time. The response can for instance be in terms of road transport tonne kilometres. If the impact of a 1% increase in the road transport cost on road freight tonne kilometres is a decrease in road tonne kilometres by 0.3%, the road transport cost elasticity of the demand for road freight tonne kilometres is -0.3 ($= -0.3/1$).

The main use of elasticities is in high-level models (e.g. de Jong, Bakker, Pieters, & Wortelboer-van Donselaar, 2004) to get a first impression of the likely impacts of some policy measure. This can be used to distinguish promising measures from non-promising ones. In a next step, the promising set is studied using more detailed models. In the high-level models, the elasticities come from specific other models, or more generally from the available literature. The more detailed models usually have no imported elasticities, but the choice behaviour represented in these models will imply certain elasticity values (that may be different for different measures, starting points and market segments in the model), that can be calculated from them.

Section 9.3 contains an exposition on how to derive elasticities from various types of transport models. The size of an elasticity value depends to a large extent on the responses of agents that are included (e.g. changes in transport efficiency, change of mode, change of supplier). A classification of these response mechanisms will be provided. Also in Section 9.3, we will review the outcomes of the elasticity literature, distinguishing transport costs and time elasticities.

Both VTT and cost and time elasiticities are also often used as a way to judge the quality of a model: does the model (on average) give values of time and time and cost impacts on transport demand that are sensible and that are in line with the literature on such values? We think this in principle is a good way to test the relative importance of factors and the response characteristics (sensitivity) of a model. But it may also lead to a phenomenon that in the meta-analysis literature is called the 'file drawer' problem: if models are judged on the basis of the existing literature VTT and elasticity values, diverging values will be regarded as a sign of error in the model and will not be accepted, used and published (but end up in the drawer). However, sometimes there might be good reasons for divergence and for change.

9.2 Freight Service Valuation

9.2.1 Use of Models Versus Calculation of Factor Cost

In the literature, two main groups of methods can be found to obtain valuations for non-monetary freight service attributes (and especially for VTT):

- factor-cost methods;
- modelling studies.

The modelling studies are discussed in Section 9.2.2; the factor-cost method is explained briefly below.

The factor-cost method takes its name from the fact that it focuses on the input factors or production factors that are needed to produce transport services. It is basically a cost-accounting method and what it does is to calculate the cost of all input factors that will be saved in case of travel time savings, or the cost of additional inputs if travel time is increased. The basic difficulty is determining which cost components should be included.

If a transport project would reduce time needed in freight transport, this could release production factors (e.g. labour, vehicles), which then could be used for other shipments. Studies that have been applying the factor method usually include labour cost among the transport time-dependent cost. These items can be calculated using data on wages paid by transport operators (and own account shippers) to transport staff. There is no consensus on the issue whether fuel cost, fixed cost of transport equipment, overheads and non-transport inventory and logistic cost should be included in the VTT. One way to investigate this issue would be the other group of methods, i.e. the modelling studies. Some researchers argue that not all labour cost should be used in the VTT, since some of the time gains cannot be used productively. Others argue that this only is a problem for the short run, and that in the long run (which is the relevant perspective in a CBA) basically all transport cost are transport time-dependent and should be included in the VTT. This issue too can be investigated by means of modelling decisions in freight transport. We will come back to these issues when discussing outcomes of factor costs and modelling studies on the VTT in Section 9.2.3.

In a model it is possible to separate out the cost related to the average transport time and the extra cost of longer than average transport time, especially of delivering too late (possibly also of delivering too early).[1] In a factor-cost calculation that uses a simple transport cost function, it is very difficult to separate out the impact of transport time variability. However, if one would use a full logistics cost function, including components for the value of the goods, deterioration of the goods and the cost of keeping a safety stock (dependent on transport time variability), both the VTT and the VTTV can be calculated as the derivatives to transport time and variability (expressed standard deviation or variance). For a further discussion of this issue we refer to Bruzelius (2001) and Vierth (2013).

9.2.2   Different Data and Discrete Choice Models for Freight Service Valuation

The modelling studies can be classified, depending on the type of data used as a basis for modelling, into:

* revealed preference (RP) studies;
* stated preference (SP) studies (also see Chapters 6 and 10).

---

[1] It must be said, however, that many models do not make a clear distinction on this.

Joint RP/SP models are also possible in freight, but have been very few so far.
RP studies in freight use data on the choices that shippers, carriers, intermediaries or drivers actually made in practice. So, the first step for an RP model is to find choice situations where these decision-makers have to trade-off time (or another freight service variable) against cost. Examples of such situations are:

- mode choice between a fast and expensive mode and a slower and cheaper mode (see Chapter 6; and for mode combined with shipment size choice, see Chapter 5);
- choice of carrier, or between own account transport and contracting out (Fridstrøm & Madslien, 1994);
- choice between a fast toll route and a congested toll-free route;
- choice of supplier (see Section 6.3.3).

After having modelled such choices, the estimated model coefficients can be used to find the freight service valuations implied by the actual choice-making outcomes.

Most existing RP freight studies that provided one or more VTTs have been based on mode choice data (e.g. road versus rail, rail versus inland waterways).

In an SP freight VTTS study, decision-makers (in practice almost exclusively shippers or carriers) are asked to elicit their preferences for hypothetical alternatives constructed by the researcher. These hypothetical alternatives refer to shipments/transports and will have different attribute levels for transport time and cost, and possibly also for other attributes of the shipment.

The setting (choice context) of the SP experiment can be that of mode choice (e.g. repeated pair-wise choices between a road and a rail alternative for the same shipment: between-mode experiment) or route choice, as in the RP. Figure 9.1 presents an example for mode choice.

Good experience in freight VTTS research, however, has been obtained in abstract time versus cost experiments in which all alternatives that are presented refer to the same mode and the same route. In an abstract time versus cost experiment the alternatives have different scores on travel time, travel cost and possibly other attributes, but the alternatives are not given a mode or route label, such as 'rail transport' of 'motorway with toll'. An example of such an abstract choice situation is given in Figure 9.2.

Which alternative would you prefer?		
	Road transport	**Rail transport**
Transport cost	710 euro	640 euro
Transport time	2 hours and 40 minutes	3 hours and 40 minutes
Delivered 20 minutes early	10%	0%
Delivered on time	70%	90%
Delivered 40 minutes late	20%	10%
	☐ prefer this road transport	☐ prefer this rail transport

Figure 9.1 Example of a Choice Situation in a Mode Choice SP Experiment.

Which alternative would you prefer?		
	Transport A	**Transport B**
Transport cost	710 euro	640 euro
Transport time	2 hours and 40 minutes	3 hours and 40 minutes
% delivered on time	90%	95%
	☐ prefer transport A	☐ prefer transport B

Figure 9.2 Example of an Abstract Choice Situation in SP.

The representation of transport time variability of a transport alternative in an SP experiment requires special attention, because it relates to a concept that many respondents find difficult to understand.

The easiest way to include variability into a transport model is to add some measure of dispersion, such as the standard deviation or variance of transport time, to the utility function that already has transport cost and time (Significance, Goudappel Coffeng, & NEA, 2012). This formulation does not require making departure time choice endogenous (which in turn requires hard-to-get information about preferred arrival times). Under certain assumptions this representation of variability is equivalent to the expected scheduling cost from scheduling theory (Fosgerau & Karlström, 2010). However, many respondents in an SP experiment on freight transport cannot be expected to understand standard deviations.

The presentation method for variability that has been used most in freight transport is the one given in Table 9.2, which presents the percentage of the goods that is delivered at the destination on time (or possibly: within a pre-specified time window). However, this does not include anything on the severity of the delays and is very hard to convert to a measure for the standard deviation (de Jong, Kouwenhoven, Kroes, Rietveld, & Warffemius, 2009).

The presentation format of variability in Figure 9.1, which has been used by Prof. David Hensher and his colleagues in Sydney goes beyond this by also presenting a shorter and a longer travel time. This can be based on an underlying standard deviation of transport time. So a model with a standard deviation coefficient could be estimated here.

In the most recent national Dutch study on the VTT and VTTV (Significance, VU University et al., 2012), variability was presented as a series of five equiprobable transport times (with five corresponding arrival times) within a single abstract transport alternative, described only verbally, not graphically (Figure 9.3). This representation was selected after having tested several formats with verbal and graphical descriptions of five transport times in a pilot, where this format was clearly best understood and preferred by respondents (Tseng, Verhoef, de Jong, Kouwenhoven, & van der Hoorn, 2009). It allows estimation of a model with variability as the standard deviation, a scheduling model and a combination of both.

Question
Which alternative do you prefer

Tropsnar A		**Tropsnar B**	
Departure time: **09:45**		Departure time: **09:20**	
You have an equal chance on each of the following transport times and corresding arrifal times		You have an equal chance on each of the following transport times and corresponiding arrival times	
Transport time:	Arrival time:	Transport time:	Arrival time:
1 h and 25 min ->	11:10	1 h and 50 min ->	11:10
1 h and 45 min ->	11:30	2 h and 10 min ->	11:30
1 h and 45 min ->	11:30	2 h and 10 min ->	11:30
2 h and 5 min ->	11:50	3 h and 10 min ->	12:30
2 h and 25 min ->	12:10	4 h and 50 min ->	14:10
Usual transport time: 1 h and 45 min.		Usual transport time: 2 hr and 10 min.	
Transport cost: € 625		Transport cost: € 625	

Prefer transport A ☐   Prefer transport B ☐

Figure 9.3 Example of an abstract choice situation in SP with variability represented in the form of five equi-probable transport times.
Source: Based on Tseng et al. (2009) and Significance, VU University et al. (2012).

SP data has some advantages in the case of freight transport modelling, in particular as it may be possible to obtain data (e.g. on costs and rates) which would be difficult to acquire by other methods (Fowkes, Nash, & Tweddle, 1991). The drawback of SP data is its hypothetical nature: these are stated responses to hypothetical choices, not actual decisions. This problem can be minimised using carefully designed SP surveys in which the respondents are asked to choose between alternatives relevant to their own circumstances (contextual stated preference). In computer-based SP experiments decision-makers, such as logistics managers, can be presented with the choice between alternatives for a specific real-world consignment. The alternatives are defined using previous answers from these respondents; the attribute levels are based on the observed levels for the selected consignment. Practically all SP surveys in freight transport have been carried out as computerised interviews, which can provide the highest degree of customisation.

A difficult issue in SP surveys on freight service valuation is who to interview on what (also see Sections 6.1 and 6.2.4). Massiani (2005) argues that shippers will only give the time value of the cargo itself (related to interest on the inventory in transit and stock-out costs), whereas the willingness-to-pay (WTP) of carriers will reflect all the components of the value of time. Booz, Allen, Hamilton & Institute for Transport Studies (2003) note that especially for carriers it might be difficult to separate between a change in time and a change in cost.

In the most recent freight VTT and VTTV study in the Netherlands (Significance, VU University et al., 2012), specific assumptions (*a priori* hypotheses) were made on the extent to which particular actors take into account different components of the freight VTT — and should do so, when responding to the SP questions (Table 9.1).

Table 9.1 Hypotheses on the Aspects that Freight Respondents Include in their
VTT (and VTTV)

	Values Related to the Cargo	Values Related to the Vehicles and Staff
Carrier	Not included	Included
Own account shipper	Included	Included
Shipper that contract out	Included	Not included

Carriers are in the best position to give the component of the VTT (and VTTV) that is related to the costs of providing transport services. If the transport time would decrease, vehicles and staff would be released for other transports, so there would be vehicle and labour cost savings.

Shippers that contract out are most interested in other aspects, as expressed by the VTT (and VTTV) that is related to the goods themselves. This includes the interest costs on the capital invested in the goods during the time that the transport takes (only important for high-value goods), the reduction in the value of perishable goods during transit, but also the possibility that the production process is disrupted by missing inputs or that customers cannot be supplied due to lack of stock. The latter two arguments are also (possibly even more so) important for the VTTV.

Shippers with own account transport can give information on both the values that are related to the costs of providing transport services and the values that are related to the goods themselves. If both these components of the VTT (VTTV) are properly distinguished, the carrier VTT (VTTV) and shipper (contract out) VTT (VTTV) can be added to obtain the overall VTT (VTTV) for use in societal CBA.

In the new Dutch study (Significance, VU University et al., 2012), VTTs and VTTVs were sought that include both components (not just the goods-related but also the services-related component), since in CBAs for transport projects in the Netherlands the user benefits of savings in vehicle and staff cost are included in the time savings of the project (unlike for instance Sweden, where the VTT only relates to the goods component, and transport cost changes are dealt with separately). Previous studies have not tried to disentangle the two VTT (VTTV) components, but this study obtained estimates for both components separately.

Of course there may be exceptions to the general pattern depicted in Table 9.3, but in the questionnaires the researchers steered the shippers that contract out only to answer on the components they generally know most about (bottom-left), and likewise for carriers (top-right). This was done by giving very explicit instructions and explanations to get clearly defined component values from each type of agent. In other words, the researchers:

- Explained to all respondents that the changes in time, costs and reliability are generic: these apply to all carriers using the same infrastructure, and are not competitive advantages for their specific firm.
- Explained to carriers (and logistics service providers) that a shorter transport time might be used for other transports: the staff and vehicles/vessels can be released for other

productive activities. A higher reliability means that the carriers can be more certain about such re-planning/rescheduling. They also explained that the carriers do not have to take into account what would happen (deterioration, disruption of production process, running out of stock, etc.) to the goods if they were late.

• Explained to the shippers that contract out that they only have to take into account what would happen (deterioration, disruption of production process, running out of stock, etc.) to the goods if the transport time or its reliability would change (whether these things would occur and how important they are was left to the respondent (shipper).
• Explained to shippers with own account transport that they have to take all of this (= cargo and vehicle) into account.

The types of models (such as multinomial logit, mixed logit) that are estimated on the types of data discussed in Section 9.2.2 and from which VTT, VTTV and other values for service quality are derived are the same as the discrete choice models discussed in Chapter 6. But for models that are built for the sole purpose of deriving monetary valuations, the mixed logit specification is nowadays more prevalent (especially in passenger transport) than among models that are estimated for use as forecasting models. This is caused by the fact that forecasting models are run many times after they have been estimated whereas for a monetary valuation model estimation is sufficient.

Furthermore, many valuation studies only use SP data. Forecasting models on the other hand are usually based on RP or joint RP/SP data (an explanation for this difference can be found in Section 6.2.2).

Many models follow a linear utility specification in time and cost:

$$U = \beta_C \cdot C + \beta_T \cdot T + \beta_R \cdot \sigma \qquad (9.1)$$

where:

U = utility
β_C = transport cost coefficient (to be estimated)
C = transport cost
β_T = transport time coefficient (to be estimated)
T = transport time
β_R = variability (reliability) coefficient (to be estimated)
σ = standard deviation of the transport time distribution.

The value of transport time VTT can be calculated by dividing the time coefficient by the cost coefficient[2] :

$$\text{VTT} = \frac{\beta_T}{\beta_C}$$

The value of variability is calculated in a similar way:

$$\text{VTTV} = \frac{\beta_R}{\beta_C}$$

[2] Exact methods also exist for calculating standard deviations or t-ratios of the VTT or VTTV on the basis of the statistics for the individual coefficients (Daly, Hess, & de Jong, 2012).

A measure that is sometimes used to express the VTTV (based on the standard deviation) relative to the VTT is the reliability ratio RR:

$$RR = \frac{\beta_R}{\beta_T}$$

Some models are not in utility space but in WTP space:

$$U = \beta_C \cdot (C + VTT \cdot T + VTTV \cdot \sigma) \tag{9.2}$$

Here the VTT and VTTV are estimated directly. The same goes for the logWTP space model (Fosgerau, 2006), where the natural logarithm is taken of the term between brackets in Eq. (9.2).

9.2.3 Outcomes for the Value of Transport Time

de Jong (2008) is a review paper on freight VTT that contains outcomes for the freight VTT for different modes from different studies reported up to 2007. In the tables below, we summarise the main findings of the 2008 paper and add some new studies. Another recent overview of VTTs is given by Feo-Valero, Garcia-Menendez, & Garrido-Hidalgo (2011).

Not all the studies included in de Jong (2008) or in the tables below were specific VTT studies; some focused on the valuation of several freight service attributes, others were designed for predicting future freight volumes. Several assumptions with regard to average shipment size, shipment value, transport cost and times had to be made and exchange rates and price index numbers were used to convert to 2010 euros. The values should therefore be only regarded as indications of the outcomes of the studies quoted. Furthermore, unlike the tables in de Jong (2008), we now tried to group the empirical outcomes (which for some studies was a somewhat subjective task) into:

*   outcomes for the goods component of the VTT;
*   outcomes for the transport service component (vehicles and staff) of the VTT;
*   outcomes for both components together.

Table 9.2 gives the outcomes for road transport. These VTTs refer to an average truck. In the Dutch freight VTT studies the average load is 8 tonnes (taking into account empty transports).

de Jong (2008) found a group of studies that obtained road VTTs in the range between 30 and 50 euro (of 2002), which in euros of 2010 would be the range between 35 and 60 euro. In the table this is repeated in the part on both components of the VTT together. Some of the studies included here are the first Dutch freight VTT study (de Jong, Gommers, & Klooster, 1992), the 1994/1995 UK VTT study (Accent & HCG, 1999), Fowkes, Firmin, Whiteing, & Tweddle (2001), the second national Dutch freight VTT study (de Jong et al., 2004) and Hensher, Puckett, & Rose (2005). The recent Norwegian freight VTT and VTTV

Table 9.2 Value of Transport time (VTT) in Goods Transport by Road (in 2010 euro per Transport per Hour)

Publication	Country	Data	Method	VTT
The goods component in the VTT:				
de Jong (2008)	Various Scandinavian studies up to 2001	SP	Different discrete choice models	0–10
Danielis, Marcucci, & Rotaris (2005)	Italy	SP	Ordered probit	7
IRE & RAPP Trans (2005), Maggi & Rudel (2008)	Switzerland	SP	MNL	14
Fries, de Jong, Patterson, & Weidmann (2010)	Switzerland	SP	Mixed logit	4
Halse et al. (2010)	Norway	SP	MNL and mixed logit	Large truck (carrying on average 12 t): 9
de Jong et al. (2011)	Netherlands	RP (mode choice)	Aggregate logit	6
Johnson & de Jong (2011)	Sweden	RP (mode and shipment size choice)	MNL and mixed logit	24
Significance, VU University et al. (2012)	Netherlands	SP	MNL	6
The transport service component in the VTT:				
Halse et al. (2010)	Norway	Cost data	Factor cost	Large truck (carrying on average 12 t): 72
de Jong et al. (2011)	Netherlands	Cost data	Factor cost	27
Significance, VU University et al. (2012)	Netherlands	SP	MNL	32
Both components in the VTT:				
de Jong (2008)	Various countries	Mostly SP	Mostly MNL	35 − 60

(Continued)

Table 9.2 (Continued)

Publication	Country	Data	Method	VTT
Halse et al. (2010)	Norway	Cost data and SP	Factor cost and MNL and mixed logit	Large truck (carrying on average 12 t): 81 Truck (carrying 8 t): 54
Significance, VU University et al. (2012)	Netherlands	SP	MNL	38

study (Halse, Samstad, Killi, Flügel, & Ramjerdi, 2010) and the new passenger and freight VTT and VTTV study in the Netherlands (Significance, VU University et al., 2012) also find values just within this range for the sum of both components.

The Norwegian study recommended using factor cost for the transport service component and model outcomes for the goods component. Model results for the transport service component were also obtained, based on the carriers. These are about 85% of the transport costs of an hour, but the authors warn that the estimate from the carriers might contain elements of the cargo component of the shippers. In the latest Dutch VTT survey, specific instructions were used to keep the cargo and transport service separate (see Section 9.2.2). Here the transport service component of the VTT is about 65% of the transport costs per hour: the carriers do not expect that time savings can fully be converted to cost savings. In the Norwegian study, the goods component for road transport is 11% of the combined VTT and in the Dutch study for road transport this is 17%. So the models estimated in both of these studies indicate that the joint VTT is somewhat (4−18%) below the factor cost, after including the goods component (which is not part of the factor cost) in the VTT. The goods component is a relative small value, which is confirmed by the other outcomes for this component from the first part of Table 9.4.

For other modes than road transport, fewer values are available from the literature. Most other TTVs refer to rail transport. Table 9.3, for rail (or combined) transport, again summarises de Jong (2008) and provides some new evidence. The outcomes in this table are expressed per tonne (for the Dutch studies 950 tonnes for a complete train was used for the average load).[3]

As for road transport, the goods component appears to be the minor component in the rail VTT. In Significance, VU University et al. (2012), the share of the goods

[3] VTT for transport by inland waterways, sea and air transport can be found in De Jong (2008) and Significance, VU University et al. (2012).

Table 9.3 Value of Transport Time (VTT) in Goods Transport by Rail (in 2010 euro per tonne per hour)

Publication	Country	Data	Method	VTT
The goods component in the VTT:				
de Jong (2008)	Various Scandinavian studies up to 2001	SP	MNL	0 – 0.1
Johnson and de Jong (2011)	Sweden	RP (mode and shipment size choice)	MNL and mixed logit	0.1
Significance, VU University et al. (2012)	Netherlands	SP	MNL	0.3
The transport service component in the VTT:				
de Jong et al. (2011)	Netherlands	Cost data	Factor cost	0.5
Significance, VU University et al. (2012)	Netherlands	SP	MNL	0.9
Both components in the VTT:				
de Jong (2008)	Various studies	Mainly SP	Mainly MNL	0.1–1.4
Significance, VU University et al. (2012)	Netherlands	SP	MNL	1.2

component in the total VTT is about 27%. The total VTT for rail per tonne is clearly lower than for road (which amounts to about 5 euro per tonne).

For passenger transport, so many VTTs are available that various meta-analysis have been carried out, that try to explain the VTT obtained from attributes of the respective countries and study methods used. For freight transport, the number of VTTs available is somewhere near the margin of what is minimally needed for a meta-regression. The HEATCO project for the EU found, in a meta-analysis explaining the freight VTT from various countries, that the GDP per capita elasticity of the freight VTT was between 0.3 and 0.4 (Bickel et al. 2006). The finding of a value clearly below unity was attributed to the openness and competitiveness of transport markets. This leads to freight rates that vary considerably less between countries than GDP per capita. Zamparini & Reggiani (2007) assembled 46 observations on the VTT in freight transport for 22 countries in Europe and North America. Their regression function explained the natural logarithm of the VTT from GDP per capita, region and mode.

9.2.4 Outcomes for the Value of Transport Time Reliability

In Table 9.4, is an overview of quantitative results for the VTTV in freight (largely based on ITS Leeds et al., 2008 and Significance, Goudappel Coffeng et al., 2012; Significance, VU University et al., 2012). As discussed in the previous section, the reliability ratio RR (that uses VTTV expressed as the standard deviation) is probably the most practical measure for including the VTTV in freight transport models. However, only few studies using this measure have been carried out. Recently, some results (Fowkes, 2006; Halse et al., 2010; Significance, VU University et al., 2012) have become available that indicate that in freight transport the RR may not be as high as previously thought (de Jong et al., 2009; MVA, 1996).

Table 9.4 VTTV in Goods Transport (in 2010 euro)

Publication	Country	Data	Method	Quantitative Outcomes (+ definition): Transport Time or Cost Equivalent
Hague Consulting Group, Rotterdam Transport Centre, and NIPO (1992)	Netherlands	SP survey among shippers and carriers	MNL	The Netherlands: an increase in the percentage not on time by 10% (e.g. from 10% to 11%) is just as bad as 5−8% higher transport costs
Accent and Hague Consulting Group (1999)	UK	SP among shippers and carriers (road)	MNL	A 1% increase in the probability of delay of 30 min or more is equivalent to 0.5−2.1 euro per transport
MVA (1996)	UK	Literature review		Reliability ratio for transport: 1.2
Small, Noland, Chu, & Lewis (1999)	USA	SP survey among hauliers	MNL scheduling model	A reduction in the deviation from the agreed delivery time (schedule delay) by 1 h is worth 450 euro per transport
Bruzelius (2001), based on Transek (1990, 1992)	Sweden	SP survey among shippers	MNL	For rail transport, a 1% increase in the frequency of delays is equivalent to 5−8 euro per wagon For road transport: 4−37 euro per transport

(Continued)

Table 9.4 (Continued)

Publication	Country	Data	Method	Quantitative Outcomes (+ definition): Transport Time or Cost Equivalent
Bruzelius (2001), based on INREGIA (2001)	Sweden	SP survey among shippers	MNL	The value of the risk of delay is 7 euro per pro mille per transport for road, 128 for rail and 30 for air transport
Fowkes et al. (2001)	UK	SP survey among shippers and carriers (road)	MNL	The value of the difference between the earliest arrival time and the departure time is on average 1.4 euro per minute per transport (more or less the free-flow time) For the time within which 98% of the deliveries takes place minus the earliest arrival time, the value is 1.7 euro ('spread') For deviations from the departure time (schedule delay) the value is 1.3 euro
de Jong et al. (2004) Also used in de Jong et al. 2009	Netherlands	SP survey among shippers and carriers	MNL	A change of 10% in the percentage not on time (e.g. from 10% to 11%) is equivalent to 2 euro per transport for road transport. When converted to reliability ratio: 1.24 Also values for rail, inland waterways, sea and air transport
Bogers & van Zuylen (2005)	Netherlands	SP among truck drivers and managers of shippers and carriers	MNL	Truck drivers value the unfavourable travel time twice as high as its objective (risk-neutral) worth. Managers of shippers and carriers did not have this relatively higher value for unfavourable travel times

(Continued)

Table 9.4 (Continued)

Publication	Country	Data	Method	Quantitative Outcomes (+ definition): Transport Time or Cost Equivalent
IRE & RAPP Trans (2005), Maggi & Rudel (2008)	Switzerland	SP among shippers	MNL	A 1% point increase (e.g. from 10 to 11%) in the percentage on time has a cost of 42 euro per shipment
Fries et al. (2010)	Switzerland	SP among shippers	Mixed logit	A 1% point increase (e.g. from 10 to 11%) in the percentage on time has a cost of 16 euro per shipment
Fowkes (2006)	UK	SP survey among shippers and carriers	MNL	Reliability ratio: 0.2–0.3
Hensher et al. (2005)	Australia	SP for tolled and toll-free roads	Mixed logit	VTTV of 2.5 euro per percentage point for transporters, 7.50 euro for shippers. This is obtained when looking solely at the freight rate; when further incorporating all costs in the calculation, the VTTV rises to 9.1 euro. Giving an actual meaning to these values, the results would imply that, if a toll-free route had a 91% probability of on-time delivery, with 97% for the tolled route, the VTTV for transporters would be 15 euro per trip
Halse et al. (2010)	Norway	SP (mainly shippers in road transport)	MNL	Reliability ratio for shippers using road transport: 1.3

(Continued)

Table 9.4 (Continued)

Publication	Country	Data	Method	Quantitative Outcomes (+ definition): Transport Time or Cost Equivalent
Significance, VU University et al. (2012)				Reliability ratio for carriers (road): 0 Overall reliability ratio for road: 0.1–0.4. Reliability ratio for shippers using road transport: 0.3–0.9 Reliability ratio for carriers (road): 0 Overall reliability ratio for road: 0.4 Also values for rail, inland waterways, sea and air transport

9.2.5   Other Freight Service Values

Other freight service quality attributes for which monetary values have been derived, mainly in SP studies, are (see Table 5 of Feo-Valero et al. (2011) for an overview by study):

• Probability of damage during transport
• Frequency
• Presence of a transhipment
• Flexibility
• Provision of information about delays
• Greenhouse-gas emissions from transport.

9.3   Freight Transport Elasticities

9.3.1   Derivation of Elasticities from Transport Models

The advantage of elasticities is that they are dimensionless, i.e. a change in the unit of measurement (for instance from kilometres to miles) does not affect the elasticities. Elasticities give the ratio of a percentage change in demand or supply (e.g. road tonne kilometres) to a percentage change in one of the factors explaining demand or supply (e.g. price of road freight transport).

In this chapter, we use the following *general definition* of elasticity:

> *An elasticity gives the impact of a change in an independent (or stimulus) variable on a dependent (or response) variable, both measured in percentage changes.*

Elasticities are defined using the *'ceteris paribus'* condition: they are valid under the assumption that all other things (e.g. other independent variables) do not change.

An elasticity can be positive or negative. If an elasticity (in absolute values) exceeds 1, the dependent variable is called 'elastic' (e.g. elastic demand) w.r.t. the independent variable.

9.3.1.1 Some Basic Distinctions

A first distinction is between point elasticities and arc elasticities. A point elasticity measures the proportionate change in the dependent variable resulting from a very small proportionate change in the independent variable. The price (P) elasticity of demand for commodity Q in terms of a point elasticity is:

$$E_p = (dQ/Q)/(dP/P) = (dQ/dP) \cdot (P/Q) \tag{9.3}$$

In this formula, dQ/dP is the derivative of the (ordinary or Marshallian[4]) demand function w.r.t. P (the slope of the demand function).

An arc elasticity is applicable if the change in the independent variable is not very small, whereas point elasticities are appropriate for small changes. An arc elasticity is defined as:

$$e_p = (\Delta Q/\Delta P) \cdot (P_1 + P_2)/(Q_1 + Q_2) \tag{9.4}$$

In which the subscripts 1 and 2 represent the situation before and after the change in price. Whether an arc elasticity will be higher or lower than a point elasticity depends on the shape of the demand function (e.g. concave or convex).

Another distinction is between own and cross elasticities. If for instance we are studying mode choice, the own (or direct) elasticity gives the impact of an attribute of some mode on the demand for that same mode, e.g. the road transport cost elasticity of road freight tonne kilometres. A cross elasticity measures the impacts on other modes, e.g. the road transport cost elasticity of rail freight tonne kilometres.

We use transport price and cost as synonymous here; most freight transport markets have small profit margins.

In Section 9.3.3, we provide some evidence from the literature on own-price elasticities. We did not include a comparison of cross elasticities, since such elasticities heavily depend on the current market shares of the mode, which can be

---

[4] Practically all elasticities in freight (and passenger) transport modelling are 'ordinary' elasticities, meaning that they contain a substitution and an income effect of a price change, as opposed to 'compensated' elasticities that keep income constant (using the Hicksian demand function). Also see Oum, Waters, & Fu (2008).

very different between different study areas (e.g. rail is used much more for freight transport in the United States than in Europe). In Section 9.3.4, we look at own and cross elasticities with respect to long distance rail transport in a specific study area (the European Union).

A disaggregate elasticity measures the reaction of an individual (can be an individual firm). Such elasticities can only be derived from disaggregate models, e.g. the (logit) mode choice models discussed below. For policy making, aggregate elasticities are mostly more interesting. They refer to the responsiveness of a group of individual firms (possibly the entire market). Aggregate elasticities can be derived from aggregate models and from disaggregate models. Elasticities that are calculated from a model depend on initial situation and/or the amount of change in the stimulus variable. In other words, the elasticity from a model can vary. There is one exception to this rule, which is called the 'constant elasticity of substitution' CES or 'double-logarithmic' function.

For instance the (own) elasticity from a logit model with a linear utility function is given by:

$$E^{ik}_{x_{rik}} = \beta_r x_{rik}(1 - P_{ik}) \tag{9.5}$$

In which:

E stands for the elasticity for the impact of a change in the rth independent variable x_{rik} that is part of the utility function for alternative i for individual (firm) k on the probability P_{ik} of k choosing alternative i.
β_r: the estimated coefficient for the rth independent variable.

Because of the presence of the P_{ik} term in Eq. (9.4), all coefficients of the model affect the elasticity, not just the one for the rth variable. The elasticity from an aggregate or disaggregate logit model (after sample enumeration) only gives the impact of the change in the independent variable on the distribution of a given total over the alternatives (such as the modal shares). This does not include an impact of the change in price or time on the total demand (overall modes), that is included in ordinary demand elasticities (Oum et al., 2008).

The double-logarithmic form is:

$$\ln(y_k) = \cdots \beta_r \ln(x_{rk}) + \cdots \tag{9.6}$$

In which:

Y_k: dependent variable (of a continuous nature), with observations $k = 1, \ldots, K$.
X_{rk}: the rth independent variable.

The elasticity for a change on x_{rk} is constant at β_r.

Elasticities usually come from models, estimated on empirical data, but in some cases, elasticities can be calculated from direct observations of the impact of a change (e.g. introduction of a toll), from before and after studies. The data used for model estimation can be time-series data, cross section data or panel data. If a time-series model contains lagged parameters, the model can distinguish between short and long term effects. Whether the effects from a cross section are short or long term depends on a judgement on the nature of the behavioural mechanisms included (e.g. location decisions are regarded as long run). In general, long run elasticities are larger than short run elasticities, because in the long run more response mechanisms are available.[5]

9.3.2 Classification of Response Mechanisms

The following response mechanisms can be distinguished (also see Table 9.5) for the example of an increase in the price of road freight transport (de Jong et al., 2010):

Table 9.5 Summary of Response Mechanisms to a Price Change

Reactions	Decision-Maker	Time Scale	Type of Effect			Dimension of Output		
			Fuel Efficiency	Transport Efficiency	Transport Volumes	Tonnes	Vkm	Tkm
1	C	S-M	X					
2	C	S	X					
3a	C	S-M		X			X	
3b	C/S	S-M		X			X	
3c	C	S-M		X			X	
3d	C/S	S-M		X			X	
3e	C	S-M		X			X	
4	C/S/R	S		X			X	X
5	S/R	S-M		X			X	
6	S	M-L			X	X	X	X
7	S	L			X	X	X	X
8a	S	L			X		X	X
8b	S	L			X		X	X
9	D	S-M			X	X	X	X

For Decision-Maker, S stands for shipper, R stands for receiver, C stands for carrier, D for consumers (Demand).
For Time Scale, S stands for short term, M stands for Medium term and L stands for Long term.
Changes in fuel prices affect fuel efficiency, transport efficiency and transport volumes. Changes in prices per vkm affect transport efficiency and transport volumes. Changes in prices per tkm affect transport volumes.

[5] This is assuming that all response mechanisms, short and long run, have the same sign. This is usually the case, but there could be exceptions, e.g. when the price of a high-capacity mode or vehicle type goes up, this could also lead to lower frequencies and bigger shipments, which by itself favours large capacity modes and vehicle types.

Changes in fuel efficiency

1. Fuel-efficient vehicles: Buy more energy-efficient trucks; in the long run, changes in fuel prices can also influence the fuel efficiency of the vehicles used (at the same transport volume), by accelerating/decelerating technological change in vehicle efficiency.
2. Fuel-efficient driving: Change in the style of driving (more energy-efficient driving).

Changes in transport efficiency

3. Load factor (the amount of goods measured in tonnes, divided by vehicle capacity): The load factor can be changed by:
   a. optimising the allocation of vehicles to shipments (e.g. acquire larger vehicles and group shipments, so that the same amount of tonnes can be transported with fewer vehicles);
   b. consolidating shipments originating from the same company;
   c. consolidating shipments originating from several companies (e.g. by doing collection rounds stopping at multiple senders, or by using a consolidation centre) and/or destined for several companies (e.g. by distribution rounds or distribution centres);
   d. changes in the number and location of depots, including consolidation and distribution centres (this can also be done by the shippers, depending on who owns these facilities);
   e. getting more return loads to reduce empty driving.
4. Change in route and time of day: This is mainly relevant for changes in prices that are differentiated by location and time of day (such as the road pricing scheme proposed for the Netherlands). But there may also be move to a more efficient route planning (e.g. fewer detours) because of the cost increase.
5. Increasing the shipment size (also implying a reduction in the delivery frequency; so this will increase inventory costs): This would be going against the trend towards more just-in-time (JIT) deliveries. Changes in road transport price might change trade-offs between transport costs and other logistics costs, such as order costs and inventory costs.

Changes in transport volumes

6. Change of mode: Substitution to and from rail, inland waterways, sea and air transport).
7. Changes in production technology (affecting the weight of the goods, e.g. trends towards lighter products).
8. Reduce transport demand in kilometres per tonne:
   a. Choice of supplier and receiver: Changes in the choice of supplier (procurement from more local suppliers, determining the origin given the destination) or in the geographical market size of the supplier (changing the destination given the origin), including changes in the degree of globalisation. This leads to changes in the origin-destination (OD) pattern of goods flows.
   b. Production volumes per location: Changes in production volumes per location, including use of raw materials and intermediate products for further processing. A producer can decide to shift its production to plants closer to its customers, to save transport costs.
9. Reduction in demand for the product.

Reactions 1–4 are decisions that are usually taken by the road haulier (carrier). The scope for doing these things depends on the current level of efficiency in logistics (which might be quite high already). Other reactions are possible for the haulier (e.g. hire cheaper foreign drivers or subcontractors; reduce other transport costs, such as fixed costs by postponing replacement of vehicles or economise on maintenance and repairs) that do not lead to a change in transport volumes.

Only when the road haulier passes on some of the cost increase to the shipper, the shipper will respond. The possibilities for passing on cost increases depend on market power, which may be different for different commodity markets (e.g. when specialized equipment is needed for transport, the hauliers might be in a better position). The response mechanisms 5−8 concern decisions that are usually at the discretion of the shipper (some decisions such as on shipment size can be taken by the sender but are more commonly determined by the receiver).

The manufacturers may pass on some of the cost increase to their clients (retailers, other producers, final consumers). This may then lead to the reduction in demand for the product (response 9).

The mechanisms 6, 7 and 9 will influence the number of tonnes transported by road transport. These mechanisms plus mechanism 3 and 5 will influence the number of vehicles used. Vehicle kilometres by road are influenced by all of these mechanisms plus the trip lengths (mechanisms 4 and 8). Tonne kilometres by road are influenced by mechanisms 4 and 6−9.

Many of these reactions (especially 7 and 8 and changes in vehicle technology) will only occur in the long run. Mechanisms 2 and 4 can be relevant in the short run and 1, 3, 5 and 9 in the short to medium run, whereas 6 is most relevant in the medium long run.

9.3.3 Outcome Range for Price (Cost) Elasticities

9.3.3.1 Road Transport

In this section, we present the key results of a literature review on own-price elasticities of road freight transport, distinguishing tonne-kilometre price elasticities, vehicle kilometre price elasticities and fuel price elasticities. Detailed outcomes, including for a segmentation of the elasticities to various types of goods, can be found in Significance & CE Delft (2010) and de Jong, Schroten, van Essen, Otten, & Bucci (2010). Changes in *tonne-kilometre prices* may result in various responses of the shipper; these were discussed in Section 9.3.2 as mechanisms 6−9. Most price elasticities in freight transport in the literature refer to changes in the price per tonne-kilometre. A change in the *vehicle-kilometre price* can have an impact on transport efficiency (mechanisms 3−5) and transport volumes (mechanisms 6−9), and can affect the output dimensions tonnes, vkm and tkm). Changes in *fuel prices* may lead to various responses of hauliers (mechanisms 1−2 for fuel efficiency and 3−5 for transport efficiency) and/or shippers (6−9).

The main conclusions from the literature review on own-price elasticities (de Jong et al., 2010; Significance & CE Delft, 2010) are summarised in Table 9.6.

Table 9.6 Results from the Literature Review on Road Own-Price Elasticities

Price Change	Impact On		
	Fuel Use	**Vehicle Kilometres**	**Tonne Kilometres**
Fuel price	− 0.2 to −0.6 33% Fuel efficiency change 33% Transport efficiency change 33% Mode and transport demand change	− 0.1 to −0.3	− 0.05 to −0.3
Vehicle kilometre price		− 0.1 to −0.8 33% Transport efficiency change 33% Mode change 33% Transport demand change	− 0.1 to −0.5
Tonne kilometre price			− 0.6 to −1.5 40% Mode change 60% Transport demand change

The above results on elasticity values are supported by almost 80% of the studies reviewed. Just above 20% of the studies yields values that are clearly lower or higher. The cell values are based on different sets of studies and not necessarily mutually consistent. A consistent set of best-guess values for road transport can be found in Significance & CE Delft (2010) and de Jong et al. (2010).

Notice that especially the values presented with regard to fuel price change are characterised by rather high uncertainties due to the limited number of studies that report estimates for these elasticities (to a lesser extent the same qualification holds for vehicle kilometre price elasticities).

9.3.3.2 Rail Transport

For rail transport, similar price changes and response mechanisms exist as for road transport. A review of the literature of rail price elasticities was carried out in VTI & Significance (2010). The literature does not distinguish all these response mechanisms separately. On the basis of the literature, the following composite response mechanisms were distinguished:

- *Change in mode*: Substitution to and from road, inland shipping and (short) sea shipping.
- *Changes in transport demand*: Due to the changes in tonne-kilometre prices shippers may choose other supplier/receivers or other production locations. These decisions may lead to changes in total transport demand (without changes in tonnes shipped).

Table 9.7 Results from the Literature Review on Rail Own-Price Elasticities

Price Change	Impact on		
	Tonnes	**Vehicle (=Train) Kilometres (vkm)**	**Tonne Kilometres (tkm)**
Price per vehicle (Train)-kilometre (vkm)	− 0.5 to −1.1 Derived from vkm price elasticity of tkm; using −0.1 for transport demand effect	− 0.9 to −1.5 Derived from vkm price elasticity of tkm; using −0.3 for transport efficiency effect	− 0.6 to −1.2 Derived from tkm price elasticity of tkm and assuming train operators internalise 30% of a trainkm price change by transport efficiency changes
Price per Tonne-kilometre (tkm)	− 0.8 to −1.6 Derived from tkm price elasticity of tkm; using −0.1 for transport demand effect	− 0.9 to −1.7 Derived from tkm price elasticity of tkm	− 0.9 to −1.7 'Recommendation' (on the basis of the literature)

• *Changes in commodity demand*: If the shippers cannot 'internalise' the transport price changes by themselves, they have to increase the price of the goods they offer. As a consequence consumer demand can fall and thereby total transport demand.

The main conclusions from the literature review on own-price elasticities for rail transport are summarised in Table 9.7. Notice that especially the values presented with regard to vehicle kilometre price change are characterised by high uncertainties due to the additional assumptions that had to be made to derive these elasticities. This table contains an internally consistent set of elasticities, because only for the bottom right cell, there was enough literature available (the values in the other cells are based on this first one). Rail operators internalise a part (here we assume: 30%; also based on Ecorys, 2005) of a rail costs increase by raising the transport (logistics) efficiency and pass the remainder on to their customers. These react to the price changes largely by adjusting the modal split, but about −0.1 of the −0.9 to −1.7 range is for changes in total transport demand (such as choosing different suppliers or customers for the commodities). These transport demand effects are considerably smaller than for road transport, since the share of rail transport in the total transport cost for all commodities is much smaller than for road transport. For the same reason we expect that there will be no change in commodity demand when rail prices change.

For practical studies that will use these above elasticities as an indication of the likely impact of a price change in rail (or road) transport, we recommend to carry out a sensitivity analysis, using different values from the range given, including the upper and lower bound.

Table 9.8 Road Transport (per tkm) Cost Direct and Cross Elasticities for Transport (in tkm) of Bulk and General Cargo, at Different Transport Distances for the EU[a] (Modal Split Effects Only)

Mode	Distance Band			
	500–1000 km		More than 1000 km	
	Bulk	General Cargo	Bulk	General Cargo
Road transport	− 0.5	− 0.7	− 1	− 0.8
Inland waterway	1	0.5	0.6	0.2
Train	1.5	1.1	1.7	1.2
Combined transport	0	1.1	0	1.2
Short sea	0.3	0.2	0.3	0.1

[a]At the time, the EU had 23 member states. This, plus Norway and Switzerland, is the study area for the EXPEDITE model.

Finally, the literature on rail price elasticities for different commodity types, distance classes and train types was analysed, in as far as available in the literature:

- Several studies where rail transport price sensitivities are larger for general cargo compared to bulk products (e.g. solid fuel, petroleum, iron ore, fertilisers, stones, wood), but some studies find the reverse.
- The price elasticities for short distance rail transport are smaller than for long distance rail transport.

9.3.4   Consistent Cost and Time Elasticities of the Modal Split

A review of freight modal split elasticities for changes in both time and cost in Europe was given in the EXPEDITE project (de Jong, 2003; de Jong et al., 2004). The outcomes of a number of national and international freight model runs were averaged in the EXPEDITE meta-model for freight transport to give consistent reactions to the different policy changes. Outliers (very high or low elasticities) were truncated, to prevent over-reaction in the meta-model, and non-availability of certain modes (e.g. sea transport in Austria) was taken into account. In Tables 9.8 and 9.9 are examples of the average elasticities from the EXPEDITE meta-model.

Increases in road transport costs for general cargo and especially for bulk goods on transport distances over 500 km have a substantial effect on modal choice. According to the meta-model, raising road transport costs will lead to a modal shift in bulk transport away from trucks mainly to rail and inland waterway transport, while for higher value general cargo the shift will mainly take place from truck to rail and intermodal combined transport. For trips up to 500 km, the elasticities are considerably smaller, between 0 and −0.3 for the road transport cost elasticity of transport of bulk products by road, and between 0 and −0.5 for general cargo.

High-value goods are usually more time-sensitive than lower value goods. In Table 9.11, general cargo deliveries show at distances over 1000 km a sharp

Table 9.9 Road Transport Time Direct and Cross Elasticities for Transport (in tkm) of Bulk and General Cargo at Different Transport Distances for the EU[a] (Modal Split Effects Only)

Mode	Distance Band			
	500–1000 km		More than 1000 km	
	Bulk	General Cargo	Bulk	General Cargo
Road transport	− 0.55	− 0.7	− 1.2	− 1.4
Inland waterway	0.8	0.4	0.5	0.15
Train	1.8	1.0	2.0	1.0
Combined transport	0	1.3	0	1.4
Short sea	0.04	0.1	0.03	0.1

[a]At the time, the EU had 23 member states. This, plus Norway and Switzerland, is the study area for the EXPEDITE model.

increase in the elasticity value. This may be explained by a critical distance at around a 1000 km. At an assumed average truck speed of 70–80 km/h, a distance of about 1000 km is an upper limit which can no longer assure timely overnight delivery. At distances below 500 km, the time elasticities of truck tkm are smaller: between 0 and −0.25 for bulk goods and between 0 and −0.5 for general cargo.

References

Accent and Hague Consulting Group. (1999). The value of travel time on UK roads. Report to DETR, Accent and Hague Consulting Group, London/The Hague.

Bickel, P., Hunt, A., de Jong, G., Laird, J., Lieb, C., Lindberg, G., et al. *HEATCO deliverable 5: Proposal for harmonised guidelines*. Stuttgart: IER.

Bogers, E. A. I., & van Zuylen, H. J. (2005). De rol van betrouwbaarheid bij routekeuze van vrachtwagenchauffeurs (The importance of reliability in route choices of truck drivers). *Tijdschrift Vervoerwetenschap, 41*(3), 26–30.

Booz Allen Hamilton and Institute for Transport Studies, University of Leeds (2003). Freight user benefits study. Assignment 01-08-66 for the Strategic Rail Authority, *Booz Allen Hamilton and ITS Leeds*.

Bruzelius, N. (2001). *The valuation of logistics improvements in CBA of transport investments − A survey*. Stockholm: SIKA (SAMPLAN).

Daly, A. J., Hess, S., & de Jong, G. C. (2012). Calculating errors for measures derived from choice modelling estimates. *Transportation Research B, 46*, 333–341.

Danielis, R., Marcucci, E., & Rotaris, L. (2005). Logistics managers' stated preferences for freight service attributes. *Transportation Research, Part E, 41*, 201–215.

de Jong G. C. (2003). Elasticities and policy impacts in freight transport in Europe. In *European transport conference*, Strasbourg.

de Jong, G. C. (2008). Value of freight travel-time savings, revised and extended chapter. In D. A. Hensher, & K. J. Button (Eds.), *Handbook of transport modelling* (1). Oxford/ Amsterdam: Elsevier Handbooks in Transport.

de Jong, G. C., Bakker, S., Pieters, M., & Wortelboer-van Donselaar, P. (2004). New values of time and reliability in freight transport in the Netherlands. In *European transport conference 2004*, Strasbourg.

de Jong, G. C., Burgess, A., Tavasszy, L., Versteegh, R., de Bok, M., & Schmorak, N. (2011). Distribution and modal split models for freight transport in the Netherlands. In *European transport conference 2011*, Glasgow.

de Jong, G. C., Gommers M. A., & Klooster J. P. G. N. (1992). Time valuation in freight transport: Method and results. In *PTRC summer annual meeting*, Manchester.

de Jong, G. C., Gunn, H. F., & Ben-Akiva, M. E. (2004). A meta-model for passenger and freight transport in Europe. *Transport Policy, 11*, 329–344.

de Jong, G. C., Kouwenhoven, M., Kroes, E. P., Rietveld, P., & Warffemius, P. (2009). Preliminary monetary values for the reliability of travel times in freight transport. *European Journal of Transport and Infrastructure Research, 9*(2), 83–99.

de Jong, G. C., Schroten, A., van Essen, H., Otten, M., & Bucci, P. (2010). The price sensitivity of road freight transport – A review of elasticities. In E. van de Voorde, & Th. Vanelslander (Eds.), *Applied transport economics, a management and policy perspective*. Antwerpen: De Boeck.

Ecorys. (2005). *Effecten Gebruiksvergoeding in Het Goederenvervoer*, Ecorys, Rotterdam.

Feo-Valero, M., Garcia-Menendez, L., & Garrido-Hidalgo, R. (2011). Valuing freight transport time using transport demand modelling: A bibliographical review. *Transport Reviews, 201*, 1–27.

Fosgerau, M. (2006). Investigating the distribution of the value of travel time savings. *Transportation Research Part B, 40*(8), 688–707.

Fosgerau, M., & Karlström, A. (2010). The value of reliability. *Transportation Research B, 44*(1), 38–49.

Fowkes, A. S. (2006). *The design and interpretation of freight stated preference experiments seeking to elicit behavioural valuations of journey attributes*. Leeds, UK: ITS, University of Leeds.

Fowkes, A. S., Firmin, P. E., Whiteing, A. E., & Tweddle, G. (2001).Freight road user valuations of three different aspects of delay. In *European transport conference*, Cambridge.

Fowkes, A. S., Nash, C. A., & Tweddle, G. (1991). Investigating the market for inter-modal freight technologies. *Transportation Research A, 25A-4*, 161–172.

Fridstrøm, L., & Madslien, A. (1994). Own account or hire freight: A stated preference analysis. In *IATBR conference*, Valle Nevado, Chile.

Fries, N., de Jong, G. C., Patterson, Z., & Weidmann, U. (2010). Shipper willingness to pay to increase environmental performance in freight transportation. *Transportation Research Record, 2168*, 33–42.

Hague Consulting Group, Rotterdam Transport Centre and NIPO (1992). De reistijdwaardering in het goederenvervoer, rapport hoofdonderzoek, Rapport 142-1 voor Rijkswaterstaat, Dienst Verkeerskunde, HCG, Den Haag.

Halse, A., Samstad H., Killi, M., Flügel, S., & Ramjerdi, F. (2010). Valuation of freight transport time and reliability (in Norwegian), TØI report 1083/2010, Oslo.

Hensher D. A., Puckett, S. M., & Rose, J. (2005). Agency decision making in freight distribution chains: Revealing a parsimonious empirical strategy from alternative behavioural structures, UGM Paper #8, Institute of Transport and Logistics, The University of Sydney.

Inregia. (2001). Tidsvärden och transportkvalitet, Inregia's Studie Av Tidsvärden Och Transportkvalitet för Godstransporter 1999, Background report of SAMPLAN 2001:1, Stockholm.

ITS Leeds (2008), Imperial College London and John Bates Services. Multimodal travel time variability, Final report, *A report for the Department of Transport*, ITS Leeds.

Johnson, D., & de Jong, G. C. (2011). Shippers' response to transport cost and time and model specification in freight mode and shipment size choice. In *second international choice modelling conference*, Leeds.

Maggi, R., & Rudel, R. (2008). The value of quality attributes in freight transport: Evidence from an SP-experiment in Switzerland. In M. E. Ben-Akiva, H. Meersman, & E. van der Voorde (Eds.), *Recent developments in transport modelling, lessons for the freight sector*. Bingley, UK: Emerald.

Massiani, J. (2005). *La valeur du temps en transport de marchandises* (Ph.D. thesis). Val de Marne: University Paris XII.

MVA (1996). Benefits of reduced travel time variability; report to DfT; MVA, London.

Oum, T. A., Waters, W. G., II, & Fu, X. (2008). Transport demand elasticities. In D. A. Hensher, & K. J. Button (Eds.), *Handbook of transport modelling* (1). Oxford/ Amsterdam: ElsevierHandbooks in Transport.

Significance and CE Delft. (2010). Price sensitivity of European road transport – Towards a better understanding of existing results. *A Report for Transport & Environment*, Significance, The Hague.

Significance, Goudappel Coffeng and NEA. (2012). Erfassung Des Indikators Zuverlässigkeit Des Verkehrsablaufs Im Bewertungsverfahren Der Bundesverkehrswegeplanung: Schlussbericht, Report for BMVBS, Significance, The Hague (see: <http://www.bmvbs.de/SharedDocs/DE/Artikel/UI/bundesverkehrswege-plan-2015-methodische-weiterentwicklung-und-forschungsvorhaben.html>).

Significance, VU University, John Bates Services, TNO, NEA, TNS NIPO and PanelClix (2012). Values of time and reliability in passenger and freight transport in the Netherlands. *Report for the Ministry of Infrastructure and the Environment*, Significance, The Hague.

Small, K. A., Noland, R. B., Chu, X., & Lewis, D. (1999). Valuation of travel-time savings and predictability in congested conditions for highway user-cost estimation. NCHRP Report 31, Transportation Research Board, National Research Council, United States.

Transek (1990). Godskunders värderingar, Banverket Rapport 9 1990:2, Transek, Solna.

Transek (1992). Godskunders transportmedelsval, VV 1992:25, Transek, Solna.

Tseng, Y. Y., Verhoef, E. T., de Jong, G. C., Kouwenhoven, M., & van der Hoorn, A. I. J. M. (2009). A pilot study into the perception of unreliability of travel times using in-depth interviews. *Journal of Choice Modelling*, 2(1), 8–28.

Università della Svizzera Italiana, Istituto di Ricerche Economiche (IRE) and Rapp Trans AG (2005). Evaluation of quality attributes in freight transport, *Research project ASTRA 2002/011 upon request of the Swiss Federal Roads Office*, Berne.

Vierth, I. (2013). Valuation of transport time savings and improved reliability in freight transport CBA. In M. E. Ben-Akiva, H. Meersman, & E. van de Voorde (Eds.), *Freight transport modelling*. Bingley: Emerald.

VTI and Significance (2010). Priselasticiteter som underlag för konsekvensanalyses av förändrade banavgifter för godstransporter, Del A av studie på uppdrag av Banverket, VTI notat 10-2010, VTI, Stockholm.

Zamparini, L., & Reggiani, A. (2007). Freight transport and the value of travel time savings: A meta-analysis of empirical studies. *Transport Reviews*, 27(5), 621–636.

10 Data Availability and Model Form

Lóránt Tavasszy[a] and Gerard de Jong[b]

[a]TNO, Delft and Delft University of Technology, The Netherlands
[b]Institute for Transport Studies, University of Leeds, UK; Significance
BV, The Hague, The Netherlands; and Centre for Transport Studies,
VTI/KTH, Stockholm, Sweden

10.1 Introduction

In this chapter, we first review the available data sources for freight transport modelling (Section 10.2).

Many data are collected by the national statistical offices (e.g. Statistics Norway). International statistical offices, such as Eurostat for the EU, and also the statistical offices at the UN, depend to a large degree on the national statistical offices of the member countries for their information.

Most of the data in freight transport are at the annual level (e.g. in tonnes transported per year). Official data sources are usually collected each year and then published in the form of yearly figures. Data on time intervals shorter than a year (weeks, quarters, months and working days) are very scarce, but some of the underlying data (trade statistics, transport statistics and traffic counts) is collected all year round and could be used (if access would be granted) to generate distributions of freight transport patterns over the year.

An impediment to detailed freight transport analysis is that some of the information, especially on individual shipments, transport cost and logistics cost, is proprietary. Firms in freight transport are usually reluctant to disclose this information to clients, competitors and the public.

In Section 10.3, we link the data sources from Section 10.2 to model components and model specifications in freight transport modelling for which these data can be used.

In Section 10.4, we further discuss the relationship between model form and data availability.

10.2 Overview of Different Data Sources for Freight Transport Modelling

10.2.1 International Trade Statistics

Trade statistics are published by the national statistical offices as well as by international organisations, such as Eurostat and the UN (e.g. Eurostat COMEXT and

Modelling Freight Transport. DOI: http://dx.doi.org/10.1016/B978-0-12-410400-6.00010-0

UN COMTRADE data). These data generally contain the amounts of import to a country and export from a country, by country, for a certain commodity classification, in monetary values (additionally the same flows measured in tonnes are often provided as well). This means that the monetary trade flow from some country A to country B (by commodity type) can be obtained both from country A (export statistics) and country B (import statistics). One of the first problems that analysts of freight data encounter is that these trade flows from different sources do not always match: country A and B record different trade flows from A to B. Even publications by international organisations may contain such inconsistencies. If one wants to use the trade data to model international freight flows, one therefore has to harmonise the trade data first (or use data that has been harmonised already, such as those provided by the ETIS(+) project for the EU: www.etisplus.eu). However, there are no commonly accepted rules (e.g. always believe the export statistics, or smoothing on the basis of time series for each source, averaging) on how to do the harmonisation.

The source for the national statistical offices for trade data are the customs authorities. Trade data are based on the documents that have to be provided in case of cross-border commodity transactions at customs. This also explains why domestic flows are not in the trade statistics. In principle there could be region-to-region trade statistics for trades within a country, but this is not registered as such (information on domestic flows can be derived from some of the sources mentioned below). The data at the customs contain more information than just the exporting and importing country, the commodity type and the monetary value. It may also contain more detailed commodity descriptions, sectors and location codes in both countries, and the shipment weight. Customs data are now often handled electronically, but we are not aware of any freight transport research project that has obtained permission to use such data (anonymised) at the micro-level. For research purposes, the customs data are only available in the form of the aggregate trade statistics released by the national statistical offices. This also means that researchers have to take as given that there is no or hardly any information on modes and transhipments (locations) in the trade statistics, no information on locations within the countries and no detail on commodities other than the classification used in the official publications. A (limited) exception to this is that the Eurostat COMEXT data contains for flows from the EU to countries outside the EU and flows from countries outside the EU to the EU information on the mode at the crossing of the EU border and whether containers were used or not.

The commodity classification commonly used in the published trade statistics is not always the same as the commodity classification that is applied in the transport statistics (that register the physical flows of goods by mode and are not based on customs data). Trade data often use the SITC (Standard International Trade Classification) classification. Transport statistics on the other hand often used the NSTR (Nomenclature uniforme des marchandises pour les Statistiques de Transport, Revisée) system of goods classification that from 2007 onwards is now being replaced in the EU by the NST2007 system of Eurostat. However, for instance for the Eurostat COMEXT and the UN COMTRADE data trade statistics in terms of the NSTR system can be provided (see www.etisplus.eu), in the form of country-to-country flows in values and tonnes.

10.2.2 National Accounts

The national accounts of some country provide a description of the flows in money units within, to and from a national economy. These are also usually collected and published by the national statistical offices.

Many parts of the national accounts are not particularly relevant for freight transport modelling, but some parts may be, especially:

Input—output (I/O) tables. These tables describe, in money units, what each sector of the economy of some country delivers to other sectors (also including final demand by consumers, import and export). An example would be the flow of ores, measured in its money value (cost or prices) from the mining sector to the iron and steel manufacturing sector. For many countries such tables exist, but only at the national level; the locations of the firms in the sectors are not recorded other than whether they are domestic or abroad.

Multiregional input—output (MRIO) tables. As the national input—output tables, this table includes the deliveries between sectors, and also contains the regions for both the producing and the consuming sector. Only few countries have such tables, and these mostly use larger zones within the country than one would want to use for a national (let alone regional) transport model. Nevertheless, MRIO tables can be a very good data source for a freight transport model, because it includes observational data on economic links between zones by sector pairs. And generally speaking it is a good strategy to start a freight model from economic linkages, because freight transport is a derived demand based on such linkages. However, if one would start freight modelling from MRIO tables, two conversions are usually necessary: (1) from a sector classification (where one uses the sector of the producer), such as SITC or NACE (Nomenclature des Activités Économiques dans la Communauté Européenne, to a commodity classification, such as NSTR; (2) from flows in money units to flows in tonnes (since that is the basic unit of most components on freight transport models). The former conversion can be done on the basis of existing standard conversion tables (which sometimes involves some approximation). The latter transformation is more difficult, and it requires value-to-weight data by commodity type, which will vary between countries, so standard tables will not be very helpful here. Data that can be used to derive value-to-weight ratios will be discussed in Section 10.3. Sometimes, national I/O tables are regionalised, using additional data on the production and consumption volumes of a sector by zone, to get a − more synthetic − MRIO table that is then used in freight transport modelling and regional economic modelling in the absence of MRIO tables from observations.

Make and use tables. These tables are related to I/O tables and also provide information about production and consumption in an economy. Make tables have commodities in the rows (using a commodity classification) and production sectors and imports in the columns. The cells then give the production of goods of each sector, in money units. Use tables have commodities (on the basis of some commodity classification) in the rows and the sectors consuming the goods (intermediate and final consumption, also investments and exports) in the columns. The cells thus give the consumption of goods of each sector, in money units. For use of these tables in freight transport modelling, the make and use tables need to be multiregional or be regionalised, similarly to the I/O tables.

10.2.3 Transport Statistics

Transport statistics such as roadside surveys providing information on vehicle origins and destinations. Whereas the trade statistics provide information on the locations of production and consumption of the goods (that can be used to build PC matrices, see Chapter 4 of this book, the transport statistics provide information about the locations where the vehicle flows started (point of loading) and ended (point of unloading). This information can be used to build OD matrices. When the transport from the producer to the consumer is a direct transport, the PC and OD flow will be the same. A difference occurs when the transport from the producer to the consumer uses a transport chain with several modes in a sequence (e.g. road first, then sea, then rail, then road). In this case one PC flow leads to multiple OD flows (in the example given to four OD flows), since the goods are unloaded and loaded (lifted) several times, unto several modes. In some cases the published transport statistics may not distinguish between different vehicle types of the same mode that are used consecutively, so that a transport chain LGV-HGV-LGV would just be one OD flow by road transport. In more detailed statistics cases these could be three OD flows.

The transport statistics consists of various parts: different modes have their own sources. The publication is usually carried out by the national statistical offices, but the data collection may originally have been done by others (e.g. ports and airports). Common characteristic is that these data are at the OD level, in terms of tonnes, the use of commodity classifications like NSTR and NTS2007 and that information on the mode is an integral part (since the data are gathered by mode). Transport statistics thus include the following mode-specific statistics:

Road transport statistics. This information needs to come from road haulage carriers (firms offering transport services by road) and shippers doing own account transport by road. Physical transport of goods (whether domestic or international) is accompanied by some paperwork required by the national authorities: consignment bills. The direct use of these bills as a research data source is discussed later, but this is not available for the national statistical offices. Therefore, they have to organise interviews with carriers and own account transports to get a picture of the transport flows by road. In the EU countries, information on origin, destination, commodity type and the load is collected from a sample of firms with trucks over 3.5 tonnes, under the responsibility of the national statistical offices, which report to Eurostat on this, following guidelines from Eurostat. This sample survey is expanded to the population and leads to published aggregate statistics on road transport volumes in tonnes (and tonne km as well as vehicle km).

The survey only deals with transports carried out by firms based in the home country (sometimes also focusing on transport by domestic firms on the national territory). In principle surveys with firms in their home country can also give information on transports in other countries. Some statistical offices have added interviews with foreign firms to their data.

The road transport statistics are generally only available in aggregate form (zone-to-zone data). In some cases it has been possible to use the underling microdata on the server of the national statistical office (e.g. in the PhD work of Abate (2014) in Denmark).

Seaport statistics. The national or international statistical offices do not publish a database of maritime flows between seaports (though there are some commercial databases with data at this level, but more often focusing on the movements of ships). They base themselves on statistics on the use of specific seaports (usually also collected by these seaports) and then publish data by port (or all the ports in a country together) on ingoing and outgoing transport, in tonnes, sometimes by country where the goods came from or went to, and by mode of appearance (containers, dry bulk, liquid bulk, roll on−roll off). For Europe, the ETISplus project has constructed maritime OD matrices on these data, also using information from the trade data (www.etisplus.eu). Port statistics often not only give tonnes but also the amount of sea containers (commonly measured in TEU - Twenty Foot Equivalent Units). The tonnes can be split over commodity types (e.g. NSTR), the TEU usually not.

Inland waterway statistics. This data mainly comes from national statistical offices (international organisations publish only very limited information on inland waterway flows), and of course only in a few countries this mode is of major relevance. The information is available at the OD level, but only provides tonnes (not containers) for a coarse classification of the commodities and of the zones abroad for international flows.

Railway statistics. Goods flows by rail are recorded by the railway companies that carry out these transports (possibly also by the clients of the railway companies) and the rail authorities record the train movements. Nevertheless, there often is hardly any information on rail freight flows available for transport research, especially now that many of the operators are private firms and are not obliged to provide information on this. As a result, information is only published for some of the years (e.g. every 5 years), and for instance the Eurostat data either has OD flows (tonnes) without commodity distinctions, or total volumes by country with a commodity classification. No comparable information on containers is available.

Airport statistics. As for seaports, the available base data are not organised in the form of OD flows, but as statistics of the incoming and outgoing flows of specific airports. The ETISplus project produced a − partly synthetic − OD matrix for air freight flows in/to/from the EU (in tonnes, no commodity distinction).

10.2.4 Shipper Surveys

Shipper surveys are interviews with shipping firms (e.g. producers of goods). Unlike the databases mentioned above, these are not data that are collected regularly by most of the national statistical offices in the world. Only some countries (mostly their national statistical offices do this) have carried out shipper surveys, and these are done with larger time intervals, not every year, sometimes on an *ad hoc* basis. The shippers are asked to provide information about a sample or their outgoing (sometimes also incoming) shipments of goods.

Well-known examples are the US Commodity Flow Survey (CFS; several years; see Vanek & Morlok, 1998), the French and Dutch shippers surveys of 1988, the Swedish CFS (2001, 2004–2005 and 2009; see SIKA, 2003), the Norwegian CFS and the French ECHO survey of 2004 (Rizet & Guilbault, 2004).

The information that is gathered includes the location and sector of the producer and consumer, the value and weight of the goods and the transport chains used (possibly multiple modes). The French ECHO survey goes beyond the other shipper surveys in that also the receiver and carrier firms involved in a number of specific shipments of the selected shippers were interviewed as well (extending it to a shipper–carrier–receiver survey). About 10,000 shipments in total could be reconstituted this way, with detailed information (modes, transhipment locations) about the different OD flows of these chains. The US and the Swedish surveys contain considerably less information per shipment, but millions of shipments in total.

Under certain conditions, some shipper surveys have been made accessible to transport researchers not only at the aggregate but also at the micro-level, for use in research projects.

10.2.5 Specific Project-Based Interview Data (Especially Stated Preference Data)

Several research projects in freight transport found that the existing data are not sufficient for their purposes and carried out their own interviews with shippers and/or carriers firms, focussing on one or more individual shipments. This happens regularly especially for projects that should provide freight values of time (or other service quality attribute values) or develop mode choice models for freight transport. The interviews can be revealed preference (observed choices), stated preference (choices between hypothetical alternatives) or a combination of both. In the former case, the survey is equivalent to a shipper survey. Examples can be found in Chapters 5 and 9.

10.2.6 Consignment Bills and RFIDs

Most of the information on individual shipments that researchers now get from shipper surveys, could also be obtained (and also for many more shipments) from the administrative documents that need to be completed for shipments (consignment bills) and from RFID (Radio Frequency Identification) tags, which are electronic tags for tracking and tracing the shipments. The consignment bills are now often completed and handled electronically and the tracking and tracing data is by nature electronic data. However, neither of this data is publicly available. For use in transport research, permission from the private firms involved would be needed. A related possibility would be if a transport researcher would be allowed to have its own (additional) tags on the shipments of a certain carrier or shipper and read out the data on where the shipments goes from this.

10.2.7 Traffic Count Data

Traffic counts in road transport can be both manual and automated (using induction loops in the road surface) counts. Both result in numbers of road vehicles on some road link that usually distinguish between trucks (buses) and cars. The induction loop data can also be used to calculate travel times, but in these data there usually is no distinction between trucks, buses and cars. Counts of trains, ships and airplanes are in principle also possible, but the collection and use of such data is uncommon outside major hubs, such as railway stations, airports or seaports. With new technology becoming available, traffic counts for all modes of transport can be based on approaches, such as satellite observation, GPS location services, traffic cameras, Bluetooth communication and cellular phones. This opens up now possibilities to create a complete picture of traffic flows in areas that were previously difficult to map.

10.2.8 Transport Safety Inspection Data

Transport safety inspectorates collect some data that might also be of use in freight transport modelling. Their main forms of inspection are usually roadside inspections and firm inspections, both checking whether working and driving time regulations and cargo weight regulations are being followed. This includes checking the working and driving times recorded by on-board units and the cargo plus vehicle weight at specific weighting sites or on the road itself (weigh-in-motion measurements).

10.2.9 Network Data

These are the standard transport engineering data on links, link capacity, nodes, distances and transport times. They can be organised by mode (road, rail, waterways, etc., networks) or be combined in a multimodal network that would also include transhipment links.

Apart from this, there can also be timetables for transport services that operate at fixed times, such as liner services in sea transport, shuttle trains, etc.

10.2.10 Cost Functions

Transport cost functions are usually given by mode, but sometimes also for different vehicle types within a mode. They might depend on the shipment size (e.g. lower unit rates for bigger shipments). The costs functions are sometimes based on data from a sample of firms (e.g. quoted freight rates or survey data on cost), but can also simply be based on assumptions provided by experts. Apart from transport costs information, logistics costs also consist of information about for instance order, storage and capital costs.

10.2.11 Terminal Data

These are data on seaports, inland ports, airports, rail terminals and consolidation and distribution centres within road transport on attributes, such as the location, types of goods, throughput and costs.

10.3 Which Data Sources Can Be Used in Which Type of Model?

In Table 10.1, we repeat the overview of data sources from Section 10.2 and add for each source its possible uses in freight transport modelling.

Table 10.1 can also be read from right to left: given that one wants to develop a certain model, base matrix or set of conversion factors in the context of freight transport modelling, the data sources that can be used for this are in the left-hand side column. This is also further discussed in the next section.

10.4 Discussion on Data Availability and Model Form

As discussed in de Jong et al. (2004, 2012), the four-step modelling structure from passenger transport has been adopted in freight transport modelling[1] with some extensions (Tavasszy et al., 2012):

- Generation models for production and attraction per sector (e.g. mining) or commodity group (e.g. petroleum);
- Distribution models, sometimes with a dependence of the distribution on the transport resistance between zones from the modal split model;
- Inventory network models as an intermediate step, between trade and transport;
- Modal split models;
- Network assignment.

However, additional steps are often needed to transform trade flows in money units to physical flows of goods in tonnes and further into vehicle flows with specific vehicle utilisation factors. These additional processes can be modelled as fixed rates, and also by explicit representation of logistics choices. Also other logistics aspects that are related to the trade-off between transport and inventory costs are

---

[1] There are also freight transport models which do not fit into the four-step model structure and which have no base in passenger transport modelling, such as models that explain the share of the monetary expenditure on a certain transport mode in total production cost, which are based on the economic theory of the firm.

Table 10.1 Different Data Sources and How They Can Be Used in Freight
Transport Modelling

Data Sources	Use in Freight Transport Modelling
Trade statistics	Estimation of PC matrices for the base year
	Aggregate gravity-type models for generation and distribution at the PC level
	Value-to-weight ratios (for exported and imported goods)
National account data	Estimation of PC matrices for the base year
	Aggregate I/O models and SCGE models for generation and distribution
Transport statistics	Estimation of OD matrices for the base year
	Estimation of gravity-type models for generation and distribution at the OD level (less appropriate than at the PC level)
	Estimation of aggregate mode choice models
	Load factors (cargo weight to vehicle capacity)
	Aggregate port choice models
	Models for road vehicle type choice, tour formation and empty driving/load factor if micro-data available
Shipper surveys	Estimation of PC matrices for the base year
	Estimation of disaggregate mode choice models
	Estimation of transport chain choice models
	Estimation of disaggregate shipment size choice models
	Estimation of disaggregate joint models (mode-shipment; mode-supplier)
	Value-to-weight ratios
Stated preference surveys	Estimation of disaggregate mode choice model
	Estimation of route choice models
	Estimation of transport chain choice models
	Estimation of disaggregate shipment size choice models
	Estimation of disaggregate joint models (mode-shipment; mode-supplier)
	Monetary value of service attributes (e.g. value of time)
Consignment bills and RFID data	Estimation of OD matrices for the base year (possibly PC, if tags stay on after transhipment or if combinations of tags registered at transhipment)
	Estimation of disaggregate mode choice models
	Estimation of disaggregate shipment size choice models
	Estimation of disaggregate joint models (mode-shipment; mode-supplier)
Traffic count data	Estimation of OD matrices for the base year
	Estimation of route choice models
	Calibration data
Traffic safety inspection data	Load factors

(*Continued*)

Table 10.1 (Continued)

Data Sources	Use in Freight Transport Modelling
Network data with costs functions	Direct input for the estimation of aggregate and disaggregate mode choice models and joint models Indirect input for aggregate distribution models Direct input for the estimation of route choice models
Terminal data	Direct input for the estimation of transport chain choice models

usually not included in freight transport models, even though the logistics solutions of firms influence the mode split.

For the generation and distribution steps at the zone-to-zone level, gravity models, I/O models and spatial computable general equilibrium (SCGE) models can be used (though especially the latter models represent much more than freight transport generation and distribution; they also have components on the labour, land and goods markets). To estimate gravity-based models, one needs trade statistics or transport statistics. Use of the former type of data gives models at the PC level, which are the zones where these goods flows are actually produced and attracted to. A gravity-type model on transport statistics is less appropriate, since for indirect (multimodal) transports, the starting and ending points from the transport data may not be the real endpoints that caused the goods flow, but just transhipment locations.

For the construction of I/O and SCGE models, one needs I/O tables and make and use tables from the national accounts data, preferably multiregional (otherwise, one might carry out a regionalisation). In order to make distribution models dependent on transport costs and time between zones, one needs to include time and distance data from transport network skims (possibly combined with cost functions), or accessibility indicators (such as logsums) from mode choice models.

There are hardly any disaggregate models for generation and distribution in freight transport. The only exception known to us as concerns a few models that handle the choice of supplier by an individual receiver, thus also including the distribution step in a disaggregate manner (e.g. Samimi, Mohammadian, & Kawamura, 2010). This may however be an attractive future extension of random utility discrete choice modelling that nowadays in freight transport is largely confined to mode choice. The downside is that such a model requires data from a shipper survey or a project-specific interview survey for estimation.

Inventory network models are relatively new (see Chapter 4). The latest experiments indicate that O/D data for trade and transport do differ to the extent that an intermediate model connecting the two is warranted. Aggregate and disaggregate models were estimated, respectively, based on publicly available statistics or on survey data. Interregional trade is not observed directly and observations of warehousing activities are only available in exceptional cases. The development of such models will be confined to situations where a suitable estimation framework can be built on the basis of available data. Supply-side data that are relevant for these models concern warehousing, handling, inventory and transport cost data for different shipment sizes and commodity groups.

The estimation of aggregate mode choice models (see Chapter 6) calls for transport statistics data by mode. These are available in most countries. The information that is required for the estimation of disaggregate mode choice models, shippers surveys, stated preference surveys or consignment bill data, is not so often available. The same data could be used for joint models of mode and shipment size (or shipment size by itself), but especially stated preference surveys are also often carried out to obtain monetary values of service attributes, such as transport time and reliability.

If the generation and distribution model would be at the PC level, the corresponding consistent mode choice component would be a transport chain choice model. In principle, this could be both an aggregate or a disaggregate transport chain choice model. However, aggregate information on actually used transport chains is very scarce and limited to records of transhipment activities at intermodal terminals and in some cases statistics of access and egress movements related to intermodal terminals. The limited direct observations that we have of transport chains comes from disaggregate shipper surveys (including CFSs). In this case, it doesn't make much sense to aggregate the disaggregate transport chain information so that an aggregate transport chain model can be estimated. Better use of the data (with less aggregation bias) then would be to use the disaggregate data to estimate disaggregate transport chain choice models (that can be used for aggregate predictions). In all cases, for a transport chain model one also needs network data (time and cost by mode), cost functions and data on the terminals for transhipment.

Information on observed use of different routes from traffic counts can be used to estimate network assignment models (alternatively in SP interviews one can collect information about stated route choices). Typically however, network assignment models do not use information on observed choices or market shares, but use a deterministic rule to assign vehicles to a shortest path (possibly also using information about link and node capacities). The only information required then is network data, in some cases with information about the value of time versus cost. In some cases traffic counts are used to calibrate parameters in the route choice model.

The same modelling philosophy that is routinely used in network assignment can also be used for other choices where information on observed choices is missing. In the absence of information about transport chains, one might use a deterministic model that predicts the transport chain choice on the basis of minimisation of the full logistics costs (Ben-Akiva & de Jong, 2013). However, the outcomes of this are normative, not necessarily realistic (the latter problem can be reduced by a calibration of the predictions to other data, such as mode shares at the OD level from transport statistics).

For building the conversion modules, data can come from trade statistics (value-to-weight ratios for export and import), and shipper surveys (export, import and domestic flows), provided that they record both the values and the weights.

Questions on the volumes (m³) might also be asked in shipper surveys and could be used as explanatory variables in mode choice and transport logistics choices (especially on the load factor). For observed information on the load factor the transport statistics (especially the OD-based interviews with the truck operators) can be used, and also the vehicle and load weight data from traffic safety inspections (though there might be a bias towards overloaded vehicles).

Aggregate port choice models can be based on port statistics (together with network information at the sea and the hinterland side). Aggregate vehicle type choice models (not ownership of the vehicles, but their use) need information on the vehicle type use shares, preferably per OD. Disaggregate vehicle type choice models are also possible, but for estimation need micro-data from interviews with the truck operators. For modelling tour formation and the amount of empty driving and the load factor of trucks at the disaggregate level, one also needs to have access to micro-data from interviews with firms that are operating the trucks.

Many transport models use the pivot-point method: the models are only used to give changes in the flows between the base year and a future year, and these changes (usually in the form of ratios, sometimes in the form of differences) are applied cell-by-cell to the base matrices (that represent the situation for the base year, as much as possible based on observed information). Pivoting can be done at the OD level as well as at the PC level (even in the same model system both can occur). PC base matrices can be based on trade statistics, national account data and shipper surveys. OD base matrices can be established on information from transport statistics, consignment bills and RFID data and traffic counts.

10.5   Dealing with Data Limitations Through Estimation

One function of transport models is to estimate missing flow data, where only a small sample of all flows can be observed, or where flows cannot be observed directly. The reasons for lack of data are often limitations in resources, different and unconnected responsibilities for data acquisition by governmental jurisdictions or different organisations, lack of sharing of data due to competition between companies. Consider for example the following cases:

- Transport databases in the form of origin/destination tables where, if no surveys are held, O/D movements are not observed directly but have to be estimated from partial observations, such as traffic counts or screenline counts.
- Transport chains, where the whole chain cannot be observed directly, unless a shipment is followed door-to-door in a transport survey. Such surveys are rare as they are very expensive. Often, however, statistics of individual modes or transhipment terminals may be sufficient to estimate the flow over individual transport chains.
- Foreign transport companies that do not have an obligation to report their movements to the statistics offices of each country of transit. In countries with a high share of foreign carriers using the road network, this implies that a significant part of the truck flows is not recorded.

- Transport terminals, where the final origin and destination of the goods transhipped is unknown but is needed to understand the size of the catchment area of terminals.
- A vehicle-based sample of O/D flows, where the share of the vehicles of the entire population of vehicles is known, but it is unclear what the share of their reported O/D movements is in the total O/D flow.

In all these cases, estimation methods can help to build a complete picture from incomplete data. O/D matrix estimation or O/D synthesis has been in use for passenger transport for decades (see e.g. Ortúzar & Willumsen, 2011). Implementations of these methods for circumstances special to freight transport date back to the 1990s (Tavasszy, 1996) and have recently been revived (Holguín-Veras, 2008). Estimation approaches can be statistical (usually based on entropy maximisation or equivalent methods) and/or based on a structural model of flows (usually the gravity models). The most basic approach towards O/D estimation is the so-called Furness algorithm. This algorithm is used to complete a matrix of flows, given (1) observed totals for columns and rows of the matrix and (2) a set of starting O/D values that provide a truthful representation of the deterrent effect of distance (or costs) on flows. It can be shown that this algorithm follows from the optimisation problem for estimating a gravity model, assuming independent Poisson distributed observations, maximising the (log)likelihood of the observed matrix totals (Kirby, 1974). The system is specified by the following equations:

$$\hat{t}_{ij} = p_i \cdot q_j \cdot f(c_{ij}) \tag{10.1}$$

$$Pr(t_{ij}/\hat{t}_{ij}) = \frac{e^{-\hat{t}_{ij}} \cdot \hat{t}_{ij}^{t_{ij}}}{t_{ij}!} \tag{10.2}$$

$$L = \prod_{ij} Pr(t_{ij}/\hat{t}_{ij}) \tag{10.3}$$

where

t_{ij} = observed transport flow between regions i and j
c_{ij} = observed transport costs
$\hat{t}_{ij}$ = calculated transport flow between i and j
L = likelihood
f = deterrence function
p, q = gravity model row and column coefficients.

The approach remains the same if the observations do not include row or column totals but parts of the matrix or sums of cells (Kirby, 1979). Depending on the type of observations that are available, additional assumptions need to be made to allow estimations. As an example, such a case occurs when, within an international O/D matrix, the transport flows between countries are not observed at a regional level, but only as country-to-country totals (Tavasszy, 1996). The estimation problem now has to satisfy two types of counts and, in addition, account for structural differences between domestic and international trade due to international border barriers. Another example involves traffic counts, which represent a sum of O/D movements. In order to know which share of movements from which O/D pairs contributes to the observed counts, we need to know the routes of all trips in the network. This can only be provided by a validated route choice model. If such a validated model is not available, the estimation of the route choice model can be taken up as part of the estimation problem. Several circumstances can complicate this basic estimation problem:

- If the route choices have a complex structure, due to e.g. round trips or hub and spoke operations, and they cannot all be modelled explicitly, an additional statistical problem needs to be solved, where the most likely trips are calculated that lead to the traffic counts observed. This problem was treated by Wang & Holguín-Veras (2009).
- If the estimation problem we have is multimodal, we need to construct a door-to-door O/D matrix from counts from different modes, connected through terminals. In this case, we need a multimodal route choice model. A model in this vein for estimating multimodal freight O/D matrices has been proposed by Pattanamekar et al. (2008).

An important precondition is that there is a minimal amount of data available, in order to avoid an underdetermined estimation problem. If matrices of observations are too sparse, independent sub-problems in the entire estimation problem can create nonsensical outcomes for the unobserved counts or the estimation algorithm will not converge (Kirby, 1979).

10.6 Concluding Remarks

Data availability is key for modelling. There is a long tradition of data acquisition for freight transport, through statistics for trade and freight trips by all modes of transport. Also, traffic counts distinguish between freight and passenger movements. At the same time, there are areas which are unobserved and notoriously difficult to map, due to the fact that current statistical systems are not developed enough, or due to the proprietary nature of business information. These include:

- Light goods vehicles, or vans, which sometimes are used for freight transport and sometimes for passenger transport. Service trips fall in between these trip purposes.
- The material content of economic exchanges, which changes due to the increasing contribution of services to economic activity, leading to dematerialisation.
- Costs of freight transport and related logistics processes (loading/unloading, cross-docking, transhipment, storage, production and administration).
- Content of transport units, be it vehicles or containers. These are observed as 'boxes' with a content that may be recorded in bills of lading but is seldom transferred into statistical systems. Trade statistics may have this detailed information but lack the specificity of transport statistics in terms of spatial detail or mode of transport.
- Consumer choices that have an indirect influence on freight but can have a strong impact, such as temporal or spatial choices in shopping behaviour, or in e-commerce.

In general, disaggregate data is hard to obtain on a systematic basis and for larger populations, without a special arrangement about confidentiality and level of detail (anonymous and generally only aggregate) in dissemination. In the past, governmental statistical offices were the ones who had exclusive infrastructures in place to create statistics for the public or to carry out unique, large-scale surveys. Currently, this field is changing quickly. First, data capture is becoming digital to an increasing extent (from paper-based surveys to web-based, further towards capture from operational transport management systems). This implies that potentially the flow of data can increase at no additional costs (or even, at lower costs). Second, as the transport world is entering the digital age, data sources are no longer isolated (by mode, firm or jurisdictional area) but can be exploited to the full as it covers entire global supply chains. This implies that transport data repositories will be created that cover entire supply chains or transport chains. Third, there is a strong drive towards sharing data across supply chains and extending the reach of data availability outside the conventional business relations to new communities. The challenge for transport modelling research is to follow these technological developments and create safe environments for experimentation, to allow development of models based on as much data as possible, and more data than was ever available. The consequences of these 'big data' developments can be huge. One could speculate that the heterogeneous nature of freight would not be an important unknown factor, as it is now in many models, but would be known in all detail. Big data analytics could also lead to the discovery of new explanatory patterns that help us to understand the drivers of freight transport demand and supply. Potentially, it could break new ground, replacing our present causal models and theories by correlations and associations between data that provide a better explanation of how freight moves.

References

Abate, A. M. (2014). Determinants of capacity utilization in road freight transport. *Journal of Transport Economics and Policy*, 48(1), pp. 137−152.

Ben-Akiva, M. E., & de Jong, G. C. (2013). The aggregate−disaggregate−aggregate (ADA) freight model system. In M. E. Ben-Akiva, H. Meersman, & E. van de Voorde (Eds.), *Recent developments in transport modelling: Lessons for the freight sector*. Bingley: Emerald.

de Jong, G. (2004). National and international freight transport models: an overview and ideas for future development. *Transport Reviews*, 24(1), 103−124.

de Jong, G., Vierth, I., Tavasszy, L. A., & Ben-Akiva, M. (2013). Recent developments in national and international freight transport models. *Transportation*, 40(2), 347−371.

Holguín-Veras, J., & Patil, G. R. (2008). Multicommodity integrated freight origin−destination synthesis model. *Networks and Spatial Economics*, 8, 309−326.

Ortúzar, J. de D., & Willumsen, L. G. (2011). *Modelling transport* (4th ed.), Chichester: John Wiley & Sons.

Samimi, A., Mohammadian, A., & Kawamura, K. (2010). Freight demand microsimulation in the U.S. In: *World conference on transport research*, Lisbon.

Rizet, C., & Guilbault, M. (2004). *Tracking along the transport chain with the shipper survey*. Proceedings of ITSC conference − Costa Rica. Elsevier.

SIKA (2003). Commodity Flow Survey 2001, Method report, SIKA Report 2003:4, SIKA, Stockholm.

Vanek, F., & Morlok, E. K. (1998). Reducing US freight energy use through commodity based analysis. *Transportation Research D, 5*(1), 11–29.

Kirby, H. R. (1974). Theoretical requirements for calibrating gravity models. *Transportation Research, 8,* 97–104.

Kirby, H. R. (1979). Partial matrix techniques. *Traffic Engineering & Control, 20,* 422–428.

Pattanamekar, P., Park, D. J., Lee, K. D., Kim, C. (2009). Estimating commodity O-D matrix using sample commodity O-D matrix from commodity flow survey and mode-specific O-D matrices. In: *88th annual transportation research board conference.* Washington, DC: TRB.

Tavasszy, L.A. (1996). Modelling European freight transport flows, PhD. Dissertation, Delft: Delft University of Technology.

Tavasszy, L. A., Ruijgrok, C. J., & Davydenko, I. (2012). Incorporating logistics in freight transport demand models: state of the art and research opportunities. *Transport Reviews, 32*(2), 203–219.

Wang, Q., & Holguín-Veras, J. (2009). Tour-based entropy maximization formulations of urban freight demand. In: *88th annual transportation research board conference,* Washington, DC: TRB.

11 Comprehensive Versus Simplified Models

Lóránt Tavasszy[a] and Gerard de Jong[b]

[a]TNO, Delft and Delft University of Technology, The Netherlands
[b]Institute for Transport Studies, University of Leeds, UK; Significance BV, The Hague, The Netherlands; and Centre for Transport Studies, VTI/KTH, Stockholm, Sweden

11.1 Introduction

The choice of the best freight transport model in a specific situation depends on many criteria, data availability (as discussed in the previous chapter) being only one of these.

The relevant criteria can be decomposed into two groups:

- The demand side: The objectives on the model and related to that the questions the model is intended to answer. But also criteria like transparency of the model for the user can be grouped here.
- The supply side: What is technically possible, including considerations of data availability, what different modelling techniques have to offer, and also the available know-how, time and money budgets for model development and run-time of the model in application?

Often different model types need to be combined in a single model system to answer specific questions. The four-stage transport models, discussed in Chapter 1, and their freight-transport-specific extensions consist of several model types (e.g. input/output (I/O) models, aggregate modal split models and network assignment) that are all needed to give the impacts on transport of adding new links to the transport networks.

A single type of model or model system that is best on all relevant criteria does not exist. Even if one would only consider the criterion of which policy questions the model should be able to answer, this would already lead to a mix of different models. The most comprehensive and complex model is not always the best model. A model should not be more complicated than is necessary to answer the questions asked (this rule is sometimes called 'Occam's razor', after the medieval philosopher who first proposed this rule). On the other hand, a model should also not be so simple that its answers will be a too inaccurate reflection of reality, which usually is very complex.

However, it may also not be wise to develop separate models for every separate policy question. Such models may be optimal on the specific criterion of providing

Modelling Freight Transport. DOI: http://dx.doi.org/10.1016/B978-0-12-410400-6.00011-2

the best possible answers to policy question, but may require much heavier invest-ments in model development than a limited number of multi-purpose models. Moreover, especially in the context of societal cost-benefit analysis (and/or multi-criteria analysis) it can be considered an advantage if several proposed transport projects and policies have been appraised using the same model, so that the out-comes will be more comparable than with different models. Multi-purpose models can also have components that can be turned on/off for answering specific questions.

So advantages of multi-purpose models are comparability of the outcomes and a more solid justification of the model development cost. But multi-purpose is not the same as all-purpose. In our view, the best choice on the criteria on model form in most situations will lead to a combination of different freight transport models for the same study area (which could be linked to each other).

In Section 11.2, we will first discuss the need to have both relatively simple models with a wide scope and comprehensive models that focus on depth of detail. Then in Section 11.3, the importance of the model objectives and the research questions on the choice of model form is discussed. The second group of criteria on model choice, the supply-side criteria, is discussed in Section 11.4. Finally in Section 11.5, we provide some concluding remarks on comprehensive versus sim-plified models.

11.2   High- and Low-Resolution Models

In de Jong, Gunn, & Walker (2004), first a review of the model types at the national and international level[1] available at the time is given, followed by a rec-ommendation to develop an integrated family of mutually consistent models at two different levels of resolution:

• a detailed, high-resolution, model system for spatial planning;
• a fast, low-resolution, policy analysis model.

The main reason for having these two different family members is that each of them can handle different questions. The low-resolution model can be used for policy analysis, which is about distinguishing between promising and unpromis-ing policy alternatives, in an uncertain world where many issues are interrelated. This should only give first order approximations, which can then be worked out into specific project proposals and subsequently be simulated in the high-resolution model to assist the actual decision-making about transport projects and policies.

Other reasons for having two sets of freight transport models at the same time for a state, country or group of countries are that the high-resolution model may be expensive and time-consuming to run for many possible policy actions, whereas

---

[1] To this family of two can be added urban freight models for cities within the national or international study area (see Chapter 8).

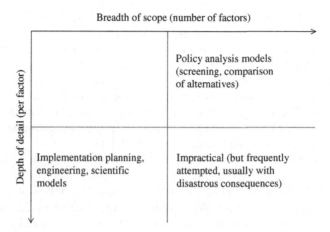

Figure 11.1 Different types of models with different scopes and levels of details.
Source: From de Jong, Gunn, & Walker (2004).

accuracy requirements and need for detail in the initial stages are lower. Finally decision-makers in different stages may have different cognitive needs and may therefore require information at different levels of detail.

Figure 11.1 shows how the low-resolution model system and the high-resolution model provide different levels for the model's scope (the breadth of the model in terms of the number of factors or markets included) and the model's depth of detail (the amount of detail for the factors that are included. Models that are neither wide nor deep are not particularly interesting. Policy analysis models (low-resolution) will preferably include a wide range of factors (e.g. not just the freight transport market but also land use, emissions and the economy), but for each of those factors limited detail will be included. High-resolution models for project appraisal and spatial planning will focus on freight transport, taking factors such as the economic conditions and land use as given (possibly through scenarios), but with more detail on freight transport itself in terms of commodity types, number of zones and size of the transport networks. Models with a lot of factors and a lot of detail per factor have also been attempted. Even though modern computing technology is able to handle much larger computational problems that in the past, 'models of everything' are not commendable. They often become highly non-transparent (the same changes can be caused by different factors) and unstable because so many things are treated as endogenous and so little is taken as exogenous.

The types of low-resolution models that come to mind for policy analysis are elasticity- and trip-rate-based models (e.g. de Jong, Gunn, & Ben-Akiva, 2004; or the HIGH TOOL model that is now being developed for the European Commission) and system dynamics models (e.g. ASTRA Consortium, 2000). Spatial computable general equilibrium (SCGE) models (see Chapter 2) also cover a considerable breadth of scope (various interconnected markets, such as for transport services, land use, labour and goods), without treating (freight) transport in detail, and might also be used for policy analysis purposes, provided that they remain relatively simple in structure and fast (and easy) in application.

A high run-time for a model is in practice often caused by equilibration processes which require that the same calculations are made over and over again to find or at least approach an equilibrium situation (iterative model applications). An example is network assignment with capacity constraints or a model with feedbacks in the form of OD transport times from assignment to transport demand. For a policy analysis model a better choice may well be to ignore such constraints and feedbacks or to approximate them within a single model run.

A low-resolution model can be developed independently, but it can also be based on one or more high-resolution models. In the latter case it becomes a 'repro-model' or 'simplified model'. One way of achieving this is to do a systematic set (but only once and for all) of runs with the detailed model, and then to estimate a repro-model on the outcomes of the detailed model, so that the low-resolution model will have basically the same response characteristics as the high-resolution model and becomes a fast and approximate version of it. One might also pull out basic equations from the detailed model and leave out equations, variables and feedbacks that are of lesser importance.

11.3   Model Objectives and Policy Questions and Their Impact on Model Form

Freight transport models are used to assess the impacts of different types of autonomous developments and policy measures, such as changes in national regulations and taxes or infrastructure investments in specific links, nodes and corridors. A wide range of models and model systems are applied by public agencies. Furthermore, a lot of freight transport modelling takes place at universities and at the individual firm level. Models to optimise transport and logistics within a specific firm or supply chain are not discussed in this chapter. Nevertheless, there are many things that models for government agencies or models in scientific research can learn from models for the private sector (as was discussed for instance in Chapters 5 and 7).

Freight transport models for public agencies are used for assisting decision-making on the following transport policy measures:

- changes in national regulations (e.g. on working and driving hours and maximum allowed vehicle loads) and taxes;
- infrastructure investments in specific links, nodes and corridors (new roads, railway lines, canals, ports, multimodal terminals, locks and also extensions of the current infrastructure in these respects);
- traffic management, such as variable message signs, on-ramp metering, variable speed limits, peak hour and reserved lanes, priorities in road and rail (e.g. freight trains versus passenger trains) traffic;
- pricing measures, such as road pricing per location and time-of-day, or railway infrastructure charges;

- spatial and temporal planning measures, such as restrictions on locations for manufacturing or warehouses, low or zero emission zones or delivery time windows for retailers.

Furthermore, there is an interest in the impact of autonomous developments (e.g. economic development, population change, employment, oil prices, . . .) on transport.

For policy questions about the influence of autonomous factors and about the impact of changes in regulations and taxes and uniform pricing measures, rather general models (like the low-resolution models discussed above) might be sufficient; detailed zoning systems and networks are not required, unless outcomes for specific zones and links would be asked.

However, for policy questions about the transport impacts of infrastructure investment projects, traffic management, charging by location and time-of-day and spatial planning measures, detailed network models are indispensable. Especially for traffic management measures, a detailed representation of the flows on the network is needed. For evaluating the impact of time-period-specific pricing measures and temporal policies, the network model needs to be supplemented by a freight transport departure time choice model (which is very uncommon in freight transport modelling, but might be done on stated preference data).

Decision-makers may want to know the impact of the above policy measures and autonomous developments (in various combinations) on transport, in the short, medium and long run, at different spatial scales. Different timescales and different spatial scales call for different types of models.

For the short run (say up to 1 year) and also the medium run (a couple of years), there is more scope for time series models, that start from the current patterns and focus on the changes over time,[2] especially if the changes are relatively small and few. For the long run (5−30 years ahead), cross-sectional models (aggregate models such as gravity or I/O models as discussed in Chapter 2; or disaggregate models such as logit models for individual mode choice as discussed in Chapter 6) that explain transport 'from scratch' may be more appropriate.

If outcomes are only required for the study area (such as a state or country) as a whole, relatively simple and fast models (such as the policy analysis models above) may be sufficient. Should outcomes be needed for a large number of zones within the study area, a high-resolution model enters the picture. An example is the appraisal of new infrastructure links, where one needs to predict an OD matrix that is assigned to the network with and without the new link to obtain the impact of the transport project on transport.

Another relevant consideration is the type of output indicators that are required. In the case of freight transport this may be (also see Section 9.3 on elasticities):

- transport volumes in tonnes and tonne km (by mode);
- vehicle km (by mode);

---

[2] This also holds for doing pivot−point analysis: this is more important for medium run forecasts than for long run predictions, since the further away one gets from the present, the less important it becomes to start from a good representation of the current patterns.

- number of vehicles on specific routes;
- number of vehicles by route and time period.

In order to get predictions for the number of tonnes and tonne km by mode one needs models of generation, spatial distribution (including inventory chains) and mode choice (or transport chain choice). But for the number of vehicle km one also needs to model the shipment size distribution, the allocation of vehicles to shipment sized and the empty backhauls (though often this is simply done by assuming fixed load factors and empty trip factors).

To generate vehicle intensities per link of the network, assignment procedures are needed. Often these are the most time-consuming parts of a model run.

Apart from the impacts of autonomous developments and policy measures on transport itself, public decision-makers often want to know the impact of these through transport on the economy and employment (the 'indirect effects of transport') and on fuel use, local and greenhouse gas emissions, safety, nature (the 'external effects of transport'). This either requires the use of unit rates for these effects (that are combined with outcomes on transport) or of specific models or model components on these issues (such as atmospheric pollution models for the spread of harmful emissions from traffic). In both cases, for CBA (Cost-Benefit Analysis) one also needs monetary values for these units.

A special kind of effect is congestion. Indicators of congestion can be the vehicle intensity to capacity (I/C) ratio of a link, the ratio of the actual to the free-flow speed or the total number of hours lost due to congestion. To obtain results in terms of these indicators, one needs to do a capacity-constrained assignment, if possible one that takes into account that from initial bottleneck links (or nodes), congestion spreads backwards through the network, affecting other links (nodes) upstream, whereas links (nodes) downstream may remain uncongested.

11.4   Approaches for Simplification

In this section, we discuss several modelling options based on our framework in Chapter 1, for simplifying high-resolution models. High-resolution models were discussed in detail in the previous chapters. The basic types of models available for low-resolution models were discussed in Section 11.2; some of these approaches also figured in earlier chapters. For more in-depth empirical information on these methods we refer the reader to Chapter 10. This section discusses the specific choice situations around high-resolution models that regularly occur in practice. We see three strategies for simplification:

- simplification by omission of sub-models;
- simplification by integration of sub-models;
- simplification by a reduced data need.

11.4.1 Simplification by Omission of Sub-Models

Figure 11.2 sketches a first series of three options for simplification of a conventional stepwise approach (numbered 1) that follows the general framework depicted in Chapter 1. The three options (left to right, numbered 2–4) are frequently encountered in practice and include the following measures:

- replacing the step of inventory networks (see Chapter 4) in the gravity model (option 2);
- replacing the I/O approach (see Chapter 2) by a direct freight generation model (see Chapter 3) (option 3);
- replacing the I/O approach by a direct trip generation model (see Chapter 3) (option 4).

We discuss these options below.

11.4.1.1 Option 2: Combining Inventory Networks and the Trade Model

This is the most frequently used approach in freight modelling. Generally, it is assumed that the trade flows in tonnes will be very close to the transport flows. Although there is usually no empirical evidence to support this assumption, it is a convenient simplification as much of the complexity of logistics can be omitted. The consequence of this simplification could be that the model underestimates the

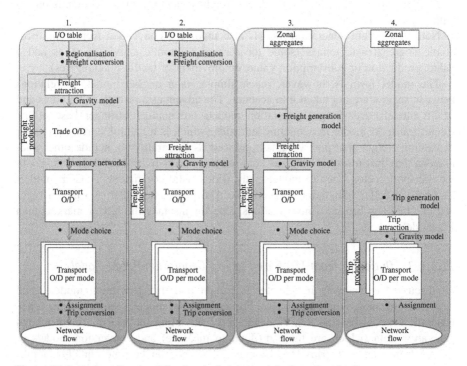

Figure 11.2 Options for simplification in freight models through reduction.

volume of flows, as indirect movements that use distribution centres are omitted. In addition, the elasticity of the transport flows will be overestimated, as inventories tend to function as buffers in the system and dampen cost increases.

A possible addition to the usual approach to take into account inventories to a limited extent is to obtain information from the I/O tables on the services provided by distribution centres, regionalise this data by the appropriate zonal statistics and use the correct conversion factors to translate these services into tonnes of freight generated or attracted. This does not guarantee yet, of course, that the spatial patterns are reproduced correctly; the gravity model is not directed at describing chains. If the amount of freight generated and attracted by distribution centres is known for each region, two gravity models can be estimated for flows to and from distribution centres. An approach for this is outlined in Davydenko & Tavasszy (2013).

11.4.1.2   Option 3: From I/O-Based to Freight Generation-Based Models

I/O and SCGE model types (see Chapter 2 or Cascetta et al. 2013) require economic statistics in the form of make/use tables or social accounting matrices. If such data would not be available (or very old) or if no reliable regionalisation (allocation of trade flows to specific geographic zones, e.g. on the basis of the share of each zone in the production and consumption of a sector) could be carried out, there is no choice really but to use direct freight generation models in combination with gravity-type models.

Note that freight generation models also involve a conversion from zonal economic aggregates to tonnes (some measure of economic activity such as added value, production or consumption value, employment or land use aggregates).

The model form is, however, much simpler than an I/O model, as relations between sectors are not taken into account. The advantages of both I/O models and SCGE models over freight generation models are a much stronger base of the freight transport model in economic statistics (which is a natural starting point for explaining freight transport volumes), as well as the ability to include other phenomena than just transport (such as land use and productivity) and their linkages to the transport sector (for I/O models only with elastic coefficients). SCGE models also have a better foundation in economic theory than the other two model types and can be even broader (more markets, also for instance the labour market) than I/O models.

11.4.1.3   Option 4: From I/O-Based to Trip Generation-Based Models

As discussed extensively in Chapter 3, freight and trip generation models each have their advantages and drawbacks. For a simplified model approach, using a trip generation model has the major advantage of obviating the use of freight-related generation data (which are more difficult to measure) and conversion models or factors from tonnes to trips (see Chapter 7), which can become equally complicated. The price is that much of the detail of the underlying logistics processes is lost, e.g. in terms of economies of density or scale that can be achieved through bundling of shipments or trips. Nevertheless, the approach is relatively easy to implement, certainly if only one mode of transport is concerned.

11.4.2   Simplification by Integration

A second strategy for simplification concerns the combination of parts of the framework into integrative models. Note that this approach, in contrast to the one above, does not eliminate parts of the framework, but mainly simplifies the *structure* of the model by combination of sub-models. Figure 11.3 shows two simplifications, one occurring in the upper third of the figure (one integrative model for the market of goods), the other in the lower third of the figure (one integrative model for transport network choice). A major advantage of integrating the production/consumption and trade sub-models for freight markets is the (theoretical and empirical)

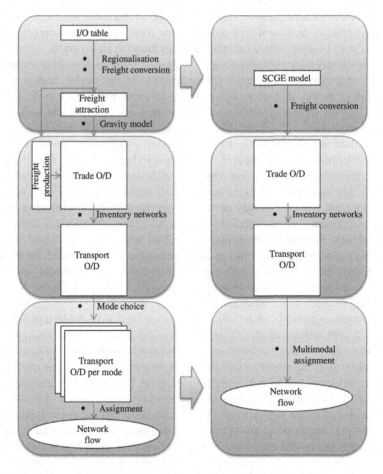

Figure 11.3 Options for simplification of the structure of freight models through combination.

consistency that is achieved between these sub-models in terms of product volumes and prices. Integrating network assignments of different modes in a supernetwork approach is useful as it provides additional information on possible intermodal transport movements. Besides this improvement in consistency and information content, the advantage of this freight model architecture is also the good fit with current policy questions in logistics (Tavasszy et al., 2003).

We will briefly discuss the pros and cons of these approaches from the viewpoint of implementation below. The SCGE approach is detailed out in Chapter 2 of this book, the multimodal network assignment in Chapter 5.

The replacement of the freight generation and distribution stages by one model, we should note, is not only reserved for the SCGE model type. Other approaches (Wegener, 2011) are possible (such as macroeconomic models, regional production function models and land use transport interaction (LUTI) models) that combine these calculations. The SCGE models, however, are rooted in one consistent body of theory (the so-called new economic geography). Nevertheless, any integrative and comprehensive treatment will require some form of equilibration (dynamic or static) and may involve longer calculation times than the base option.

Multimodal network modelling (see Chapter 6) requires less data on observed transport outcomes than aggregate choice models. In the model, transport chains with different modes in a sequence and transhipment locations can be found by searching for the shortest (fastest or cheapest) path in a multimodal network, and all that is required is this multimodal network. For validation purposes, however, additional data is required as the model generates transhipment flows. The downside of a deterministic assignment is that the researcher has little scope for controlling this optimisation process (e.g. through calibration parameters), because there are hardly any such parameters. In reality mode-route alternatives may be chosen in quite different proportions than obtained from the costs minimisation in the multimodal assignment, because decision-makers also take other factors into account (e.g. reliability, flexibility, perceptions on certain modes). In stochastic (e.g. random utility) models of mode choice such influences are accounted for in modal constants and error terms.[3] Furthermore, deterministic multimodal assignment might lead to overreactions to exogenous changes, because of the all-or-nothing character of the underlying mechanism.

Our recommendation is to handle mode choice, and if possible transport chain choice in a probabilistic model. This can either be a probabilistic discrete choice model (aggregate or disaggregate) or a probabilistic multimodal assignment (all these models were discussed in Chapter 6). If one would include the mode choice in a larger model system as a discrete choice model, the subsequent assignment can be uni-modal. In case of a discrete choice transport chain model, the assignment still needs to determine the optimal transhipment locations for every type of transport chain (e.g. which ports are optimal for road—sea—road?), as well as the best route for each uni-modal leg of the transport chain (two road legs and one sea leg in the example just given). Including all of this in a discrete choice model would

---

[3] In some stated preference models these factors have been made explicit as attributes of the modes.

lead to an abundance of choice alternatives (with mutual correlations). A choice model for network assignment that deals with this additional complexity of route overlaps is C-logit.

11.4.3 Simplification by Reduced Data Need

A third strategy for model simplification concerns the reduction of the specification of sub-models (and, in particular, the choice models) by using aggregate instead of disaggregate data. We explore this strategy for the choice model where these choices have been most debated: the mode choice model.

Aggregate modal split models (see Chapter 6) require for estimation only data on the shares of the mode by OD or PC pair (combined with cost and/or time by mode), if possible by commodity type. For disaggregate models, micro-data about the mode choice for specific shipments are needed. Disaggregate models have as advantages that they have a more direct base in a theory of individual or company behaviour and that it becomes possible in these models to include more attributes, such as those related to the shipper, the receiver, the carrier or the shipment as explanatory variables in the model. The main advantage, however, is that they do not assume that there is an optimisation of mode choice at the zone-to-zone level, but at the level of individual shipments (though possibly allowing for consolidation of individual shipments).

So, if a sufficiently large sample of micro-data on individual shipments is available, it remains hard to argue in favour of aggregate models, and the researcher is recommended to treat mode (or transport chain) choice in a disaggregate fashion. In the absence of such data, there are still possibilities for developing a deterministic micro-level model, but this would be lacking a direct empirical basis. An aggregate modal split model would be a perfectly justifiable choice under such circumstances.

11.5   Concluding Remarks on Comprehensive Versus Simplified Models

Our preferred answer to the question whether one should have a comprehensive or a simplified model is to have both types of models. The simplified model can be used for initial screening of policy options and projects and for the impact of more general (not location- and time-specific) measures. The comprehensive model then is the most appropriate model to use for assisting project appraisal, traffic management and policy measures that are location- and/or time-specific.

The choice of model type in specific situations (e.g. choice of a generation/distribution model or choice of a modal split model) not only depends on data availability but also on theoretical considerations, the question how many and which explanatory variables one wants to include and the question whether one wants to represent links with other sectors (e.g. the wider economy) or not.

References

ASTRA Consortium (2000). ASTRA final report: assessment of transport strategies. University of Karlsruhe.

Cascetta, E., Marzano, V., Papola, A., & Vitillo, R. (2013). A multimodal elastic trade coefficient MRIO model for freight demand in Europe,. In M. E. Ben-Akiva, H. Meersman, & E. van de Voorde (Eds.), *Freight transport modelling*. Bingley: Emerald.

Davydenko, I., & Tavasszy, L. A. (2013), Estimation of warehouse throughput in a freight transport demand model for the Netherlands. In: *92nd annual meeting of the transportation research board*, Washington DC.

de Jong, G. C, Gunn, H. F, & Ben-Akiva, M. E (2004). A meta-model for passenger and freight transport in Europe. *Transport Policy, 11*, 329−344.

de Jong, G. C, Gunn, H. F, & Walker, W (2004). National and international freight transport models: overview and ideas for further development. *Transport Reviews, 24*(1), 103−124.

Tavasszy, L. A., Ruijgrok, C. J., & Thissen, M. J. P. M. (2003). Emerging global logistics networks: implications for transport systems and policies, *Growth and Change: A Journal of Urban and Regional Policy, 34* (4), 456−472.

Wegener, M. (2011). Transport in spatial models of economic development. In A. de Palma, R. Lindsey, E. Quinet, & R. Vickerman (Eds.), *Handbook in transporteconomics* (pp. 46−66). Northampton: Edward Elgar.

Printed in the United States
By Bookmasters